CHANGING GEOGRAPHY

SERIES EDITOR: **JOHN BALE**

Wolfreton School

This book is to be returned on or before the last date stamped below

...e and nation

DAVID STOREY

Geographical Association

ACKNOWLEDGEMENTS

I would like to thank friends, colleagues and students at University College Worcester for providing an enjoyable and supportive atmosphere in which to work. I am particularly grateful to Claire Palmer for proofreading the manuscript. I would also like to thank the series editor John Bale for inviting me to contribute to this series, and the editorial staff at the Geographical Association for their guidance. Diane Wright's suggestions, particularly in relation to the activity boxes, were greatly appreciated.

AUTHOR: Dr David Storey is Senior Lecturer in the Geography Department at University College Worcester.

© David Storey, 2003

This book is copyright under the Berne Convention. All rights are reserved. Apart from any fair dealing for the purpose of private study, research, criticism or review, as permitted under the Copyright, Designs and Patents Act 1988, no part of this publication may be reproduced, stored in a retrieval system, or transmitted in any form or by any means, electronic, electrical, chemical, mechanical, optical, photocopying, recording or otherwise, without the prior written permission of the copyright owner. Enquiries should be addressed to the Geographical Association. The author has licensed the Geographical Association to allow, as a benefit of membership, GA members to reproduce material for their own internal school/departmental use, provided that the author holds copyright. The views expressed in this publication are those of the author and do not necessarily represent those of the Geographical Association.

ISBN 1 84377 094 6
First published 2003
Impression number 10 9 8 7 6 5 4 3 2 1
Year 2006 2005 2004

Published by the Geographical Association, 160 Solly Street, Sheffield S1 4BF. The Geographical Association is a registered charity: no 313129.

The Publications Officer of the GA would be happy to hear from other potential authors who have ideas for geography books. You may contact the Officer via the GA at the address above.

Edited by Rose Pipes
Designed by Arkima Ltd, Dewsbury
Printed and bound in Spain by EspaceGrafic

CONTENTS

		EDITOR'S PREFACE	4
		INTRODUCTION	5
CHAPTER	1.	CITIZENSHIP, INCLUSION AND EXCLUSION	7
	2.	THE STATE	15
	3.	THE NATION	26
	4.	GLOBALISATION, STATE AND NATION	34
	5.	GLOBAL WORLD, GLOBAL CITIZENSHIP	40
		REFERENCES AND FURTHER READING	47

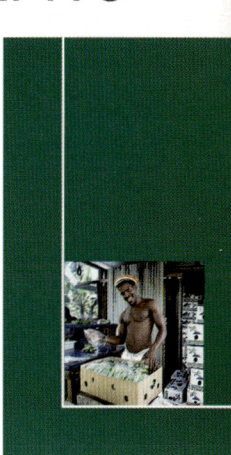

EDITOR'S PREFACE

The books in the *Changing Geography* series seek to alert students in schools and colleges to current developments in university geography. The series also aims to close the gap between school and university geography. This is not a knee-jerk response – that school and college geography should be necessarily a watered-down version of higher education approaches – but as a deeper recognition that students in post-16 education should be exposed to the ideas currently being taught and researched in universities. Many such ideas are of interest to young people and relevant to their lives (and school examinations).

The series introduces post-16 students to concepts and ideas that tend to be excluded from conventional school texts. Written in language which is readily accessible, illustrated with contemporary case studies, including numerous suggestions for discussion, projects and fieldwork, and lavishly illustrated, the books in this series put post-16 geography in the realm of modern geographical thinking.

There has been much debate in recent years over the importance of citizenship. This book examines citizenship in both the traditional sense of the relationship of the individual to the state and in the broader sense of our relationship to wider society both nationally and globally. It explores what this might mean in an increasingly globalised world. In doing so the book touches on ideas of the state, the concept of national identity and globalisation. Issues of culture, belonging, inclusion and exclusion underpin the discussion.

John Bale
February 2003

INTRODUCTION

Citizenship refers to the relationship between individuals or groups and the wider society in which they live. Traditionally citizenship has been viewed in terms of 'rights' to which people are entitled and sets of responsibilities that they have towards society. In return for discharging their civic duties people are granted certain rights.

In essence citizenship can be likened to membership of a community. Usually this is seen in national terms and most discussions of citizenship are set within the context of the relationship between individuals and the state. People are seen as 'belonging' to a particular country. This is usually reflected in the possession of a passport (or the right to possess one) of the country of which you are a citizen.

Conventionally, rights refer to things which it is generally assumed people are entitled to such as freedom of movement, free speech, an entitlement to vote and so on. Duties refer to those things required of people in order to retain full citizenship. These would include obeying the law, or generally behaving in a way not seen to threaten the stability of society.

While citizenship has traditionally been conceived as the relationship between people and the state, this idea has broadened in recent years to encompass our relationships as individuals with our local communities and with the world beyond our own country. What are our responsibilities towards those who live beyond our borders? Another facet of this broadening conceptualisation of citizenship is a concern with the environment. What are our responsibilities towards protecting and conserving the environment and building for a sustainable future? It can be argued that we have responsibilities towards future generations, not just towards the present.

In this book an introduction to the concept of citizenship is followed by an outline of the nature and role of the state; the political institution of which citizens are deemed to be members. This feeds into a discussion of the nation; the idea of a broad political community which is closely linked to the concept of the state and which is of considerable importance in making people aware of their role as citizens. Following from this the pressures currently faced by states in an increasingly globalised world are explored along with their implications for traditional ideas of citizenship within the boundaries of the state. Finally, a vision of trans-national citizenship in a more global era is elaborated upon.

© Timothy Holt.

CHAPTER 1

CITIZENSHIP, INCLUSION AND EXCLUSION

Citizens, rights and duties

Although the concept of citizenship has evolved over time, many of our ideas on the nature of it stem from the philosophers Thomas Hobbes (1588-1679) and Jean Jacques Rousseau (1712-78). The former saw the relationship between people and the state as one where individuals subordinated themselves to a state and allowed that state (or its government) to act on their behalf. Rousseau enunciated the idea of the 'social contract' whereby people gave their consent to be governed in return for the protection of the monarch or an equivalent form of government (see Faulks, 2000). There is obviously a trade-off here. Citizens will be protected by the state in return for giving their allegiance to it. It is this balance which permits societies to reduce a person's rights if that person is found to have failed in their obligations. Thus, people who commit crimes are seen to transgress that boundary by breaching the laws of society and the sanction that is imposed may, depending on the nature of the crime, result in a reduction of their citizenship rights. For example, if a person is sent to jail then his or her freedom is obviously curtailed. He or she no longer has freedom of movement.

One of the clearest statements of the modern concept of citizenship is that made by T.H. Marshall (1950). He distinguished between three different, though related, sets of rights.

1. **Civil rights**: freedom of speech, travel, etc.
2. **Social rights**: basic standard of living, adequate health care, etc.
3. **Political rights**: right to vote, engage in political activity.

Many countries outline formally the rights their citizens can expect. Possibly the best known statement of international rights is the *Universal Declaration of*

Information Box 1: UN *Universal Declaration of Human Rights*

The UN *Declaration* (adopted and proclaimed in 1948) states that all human beings are born free and equal in dignity and rights regardless of race, colour, sex, language, religion, political or other opinion, national or social origin, property, birth or other status.

The *Declaration* (which can be viewed on the UN website) covers a range of specified rights, which include:

- life, liberty and security of person;
- freedom from torture, cruel, inhuman or degrading treatment or punishment;
- equality before the law and equal protection of the law;
- freedom from arbitrary arrest, detention or exile;
- to be presumed innocent until proven guilty;
- freedom from arbitrary interference with personal privacy, family, home and from attacks upon one's honour;
- freedom of movement and residence within the borders of each state and the right to leave any country, including one's own, and the right to return to their country;
- political asylum;
- freedom of thought, conscience and religion;
- freedom of opinion and expression;
- freedom of peaceful assembly and association;
- employment, good working conditions and protection against unemployment;
- equal pay for equal work;
- freedom to form and join trade unions;
- an adequate standard of living;
- education;
- to participate in the cultural life of the community.

CHAPTER 1: CITIZENSHIP, INCLUSION AND EXCLUSION

Human Rights produced by the United Nations (UN, 1948). Through the *Declaration*, the UN endeavoured to produce a common set of rights to which all countries should adhere (Information Box 1 and Activity Box 1).

A quick glance through the UN list should be enough to make us realise that throughout the world many of these rights are not granted and that, even where they *are* granted, they may be quite fragile and subject to withdrawal. It should also be obvious that these rights are not simply given; they usually have to be struggled for. The right of women to vote in the UK and elsewhere came about after prolonged protests and campaigning by the suffragette movement. Open elections were not the order of the day in the Soviet Union or eastern Europe during the Cold War period (from the late 1940s through to the late 1980s) (see pages 19-20). The people of East Timor have recently (19 May 2002, see also Figure 19, page 29) achieved national recognition after a bloody and prolonged battle with Indonesia, which was largely ignored in Europe.

Geographies of exclusion

Leaving aside the geographically uneven nature of human rights, a discussion of citizenship automatically links into ideas of exclusion and inclusion. Those who are accepted as full citizens within a society can be deemed to be 'in'; others, however, are considered 'out'. Because of this it is important to consider the extent to which people may be excluded from full citizenship. This can occur either formally or informally – both of which can be investigated in number of ways.

Formal exclusion

Historically women were not deemed to be full citizens in many countries. Among other things they

Figure 1: In 1908 suffragettes demonstrated for women's rights. Photo: Mary Evans/The Women's Library.

were not allowed to vote. In Britain this right was obtained in 1928 only after prolonged public protest (Figure 1). The right of women to vote was denied in some other countries well into the twentieth century. For example, in Switzerland women could not vote in federal elections until 1971, thus women were effectively not deemed to be full citizens.

Usually, people born in a country are entitled to citizenship of that country (though there are many exceptions to this rule). Immigrants to many countries are not normally entitled to full citizenship but if they fulfil certain requirements they may be able to become citizens – a process frequently referred to as 'naturalisation'. For a foreign national to become a citizen of certain countries it may be necessary to:

- live there for a certain minimum period of time;
- be married to a citizen of that country;
- be able to speak the language;
- conform with other cultural criteria.

Activity Box 1: Basic human rights

In groups, consider the following points and discuss your responses to each one:

1. Using the UN *Declaration* list (Information Box 1) as a guide, what do you consider to be basic human rights?
2. Identify any rights that you believe may conflict with each other.
3. Consider the ways in which human rights appear to be both time-specific and place-specific.
4. Why might ideas of human rights change over time or between different countries?

You may find it useful to debate basic human rights further with the rest of the class.

Bearing in mind the UK and US citizenship criteria shown in Information Boxes 2 and 3, discuss as a whole class whether you regard some of these criteria as unfair or discriminatory, and if so, in what ways.

CHAPTER 1: CITIZENSHIP, INCLUSION AND EXCLUSION

Information Box 2: UK citizenship requirements

In order to obtain British citizenship, the Immigration and Naturalisation Directorate (see Websites, page 48) states that foreign nationals must:

- have lived legally in the United Kingdom for five years;
- be 18 or over;
- not be of unsound mind;
- be of good character;
- have sufficient knowledge of English, Welsh or Scottish Gaelic (depending on their age and physical and mental condition); and
- stay closely connected with the United Kingdom.

The wife or husband of a British citizen may apply for naturalisation after living in the United Kingdom legally for three years.

Information Box 3: US citizenship requirements

The US citizenship (Immigration and Naturalization Service, see Websites, page 48) requirements can be summarised as follows:

- Applicants must be at least 18 years old.
- An applicant must have been lawfully admitted to the United States for permanent residence.
- They must have lived in the US for at least five years.
- They must show that they are 'of good moral character'.
- An applicant must show that he or she is attached to the principles of the Constitution of the United States. As proof of this they must swear the oath of allegiance.

> **United States oath of allegiance**
> 'I hereby declare, on oath, that I absolutely and entirely renounce and abjure all allegiance and fidelity to any foreign prince, potentate, state, or sovereignty of whom or which I have heretofore been a subject or citizen; that I will support and defend the Constitution and laws of the United States of America against all enemies, foreign and domestic; that I will bear true faith and allegiance to the same; that I will bear arms on behalf of the United States when required by the law; that I will perform non-combatant service in the Armed Forces of the United States when required by the law; that I will perform work of national importance under civilian direction when required by the law; and that I take this obligation freely without any mental reservation or purpose of evasion; so help me God.'

- Applicants must be able to read, write, speak, and understand words in ordinary use in the English language.
- An applicant must demonstrate a knowledge and understanding of the history and government of the United States.
- A person is ineligible if they have ever been convicted of murder or carried out any of the following:
  - crimes involving moral turpitude;
  - two or more gambling offences;
  - been involved in prostitution or commercialised vice;
  - been involved in smuggling illegal aliens into the United States;
  - is or has been a habitual drunkard;
  - is practising or has practised polygamy.

Examples of the criteria for obtaining United Kingdom and United States citizenship are provided in Information Boxes 2 and 3.

Even though a person may reside in a particular country, it may be the case that they cannot obtain full citizenship. A number of consequences may flow from this, including, for example, a prohibition on voting. There may also be other difficulties for people in this situation. For example, there may be restrictions on their ability to work, to access housing or welfare services, and so on.

'Non-citizens': the case of asylum seekers

A contemporary example of this idea of formal exclusion is the case of refugees and those seeking political asylum. Within recent years the issue of refugees has assumed prominence throughout western Europe. In part this has been related to the increased numbers of people seeking asylum in European Union (EU) countries as a consequence of war and political turmoil in places such as Bosnia-Herzegovina, Kosovo and Afghanistan, and those escaping political oppression in countries such as Nigeria and Iraq (Figure 2, see Activity Box 2 and Amnesty International, 1997). In response to this the European Union has endeavoured to move towards a harmonisation of its immigration policies. These moves have been criticised by some because they make it more difficult for many people, particularly those from African and Asian countries, to enter this relatively affluent part of the world. This is in marked contrast to the historical colonisation of much of Africa, Asia and the Americas by European settlers during the colonial period when European powers tended to act as though they had an automatic right not only to reside in those places, but also to control them. It might also be argued that such moves could be interpreted as breaching the UN *Universal Declaration of Human Rights* (Information Box 1, page 7), which supports the granting of asylum to those fleeing various forms of persecution.

Some people in the UK have grown increasingly concerned over what they see as a 'flood' of immigrants. Various politicians and journalists have suggested that many asylum seekers are 'bogus' and are really only seeking economic betterment (as though there was something reprehensible about doing this). As a consequence, the UK now has pretty tough measures which discourage people from travelling to it. These include penalties on transport companies found to have immigrants without proper documentation on board, and the dispersal of asylum seekers throughout the United Kingdom, to prevent their clustering in London and the south-east. Until 2002 vouchers (rather than cash) were issued to asylum seekers. These could be exchanged in shops for food and other items. In addition, asylum seekers have not been allowed to work while their claim is being processed.

Someone applying for asylum can be refused, given permission to stay, or, in some instances, allowed to remain for a specified period of time. Recent figures on asylum applications to the UK are given in Table 1 while Table 2 indicates the main countries from which asylum seekers originate.

Figure 2: 'Another border' shows Rwandan refugees.
Photo: Howard A. Davies/Exile Images.

Table 1: Applications for asylum in the UK, 2000-02. Source: Home Office website

Time period	Number of applications
January-March 2000	18,900
April-June 2000	20,125
July-September 2000	20,435
October-December 2000	20,855
January-March 2001	18,905
April-June 2001	15,895
July-September 2001	18,860
October-December 2001	17,705
January-March 2002	19,520
April-June 2002	20,400
July-September 2002	22,560
October-December 2002	23,385

Table 2: Nationalities of asylum applicants, UK, October–December 2002. Source: Home Office website

Country of origin	Number of applications
Iraq	4375
Zimbabwe	2750
Somalia	1835
Afghanistan	1350
China	905
Iran	830
Turkey	815
Democratic Republic of Congo	705
Pakistan	565
Jamaica	565
Others	8690

Informal exclusion

The discussion above refers to people being formally excluded from full citizenship. However, even where people are nominally full citizens, this does not prevent them from being treated as second-class citizens. Minority groups such as people from certain ethnic backgrounds, gays or lesbians, people with disabilities, and women (who are in fact the majority – 51-52% of the world's population), may often perceive their treatment as being different from the norm, thereby excluding them from supposedly 'normal' society (Sibley, 1995).

As mentioned earlier, women obtained the right to vote in the UK early in the twentieth century. Despite this, more subtle forms of exclusion have continued to operate. It is only relatively recently that the idea of equal pay for women doing similar jobs to men has been considered an important issue. Indeed, some people still believe that a woman's place is in the home or in the kitchen. In other words, women's 'place' is defined in relation to their gender. Although in recent decades considerable progress has been made with regard to the status of women in society, they are still quite often faced with discrimination (sometimes subtle rather than overt) and, as a consequence, must contend with attitudes which deny them true equality. Through explorations of the nature of gender relations and gender inequality, feminist geographers have played a role in highlighting the many ways in which women may

Activity Box 2: Seeking asylum

Read the following extract. It is taken from a letter written by an Afghan asylum seeker and appeared in *The Observer* newspaper on 2 June 2002.

'I am an Afghan asylum seeker and have been waiting for a decision from the Home Office for two years … Before coming here my parents and one of my brothers were killed. I was jailed, tortured and beaten for eight months. I lost contact with almost all my friends and family. I was jailed in a place where even animals would not like to sleep. My crime was that I cried for freedom. So did my parents. I struggled and wrote against inhumanity in my country. I established an education centre … I wrote for a weekly publication which was fighting for democracy and freedom. In some of my articles I tried to write about Afghanistan, to show to the young generation who was responsible for our tragedy. The problem is that the same guys who wanted to kill me because I tried to show their faces, still rule my country.'

Now examine the websites of the UN High Commission for Refugees and the UK Refugee Council (see 'Useful websites' on page 48). Using all of this information consider the following questions:

1. Do you think the UK should accept more asylum seekers?
2. Why do you think this?
3. Do you feel people should be allowed to move freely between countries?
4. Do you feel you should be allowed to move freely between countries, work where you choose, etc.?
5. Do you feel the UK regulations on asylum applications are too generous or too restrictive?

Think about these issues individually and then discuss your responses in groups of four or five. In your discussions, try and envisage how you would respond to these questions if you were an asylum seeker. Would you answer differently? Each group can then present their findings to the rest of the class.

CHAPTER 1: CITIZENSHIP, INCLUSION AND EXCLUSION

Activity Box 3: Investigating equal opportunities

Examine the following equal opportunities statement for University College Worcester (provided by the Personnel Department), and if possible refer to other examples (your school or college should have one).

'The College is committed to implement equality of opportunity and is opposed to all forms of discriminatory practices and attitudes. The College will function in such a way that it does not discriminate unfairly – directly or indirectly – in the appointment, development, and promotion of staff on the grounds of gender, race, disability, colour, sexuality, age, nationality, ethnic or national origins, marital status, family or other care responsibility, socio-economic background, trade union activity, political or religious belief.'

Draft an equal opportunities policy for your local youth club, community group or similar organisation. In your policy statement you should:
- lay out clear objectives, and
- outline specific measures to ensure that those objectives are achieved.

be excluded from certain places and spaces (see, for example, McDowell and Sharp, 1997).

Similar issues arise in relation to race, ethnicity, disability and sexual orientation. Even though many countries, including the UK, have legislation that effectively makes it illegal to discriminate on these bases, it does not eradicate the racism, disablism, sexism or homophobia which some people express, and which serve to make members of specific groups feel like second-class citizens. They are not fully included within society. Again, geographers concerned with issues of race and sexuality have played a role in highlighting inequalities and processes of exclusion and the myriad ways in which people seen to be minorities are treated as less than first-class citizens (Jackson, 1989; Bell and Valentine, 1995). In *Disability, space and society* Rob Kitchin (2000) focuses on the ways in which people with disabilities are often excluded from particular spaces. One reaction to these issues has been the introduction of legislation that encourages employers and organisations to adopt equal opportunities policies in order to prevent discrimination against these sometimes excluded 'others' (Activity Box 3).

Geographies of power

The above discussion clearly shows how processes of exclusion operate and that it is impossible to consider ideas of citizenship without considering power. However, the issue of power is often overlooked. In focusing on the rights and duties which people have towards society it is sometimes forgotten that people need to be empowered with the ability actively

Information Box 4: The *roma*

An example of a particular group of people who are often not regarded as full citizens are the *roma* or gypsy peoples. Traditionally they were nomadic and their origins are the subject of debate (Fonseca, 1996), but, partly as a consequence of their nomadic history, *roma* have been the victims of intense discrimination in many countries.

Roma peoples are quite numerous in certain parts of Europe. With the collapse of communism in the late 1980s, the position of *roma* in eastern European countries became increasingly problematic and many were subject to various forms of discrimination and oppression. Following its independence in 1991, the Czech Republic refused to grant automatic citizenship to those people born outside its new borders. The *roma* suffered from this ruling because many of the Czech Republic *roma* were born in Slovakia (the other 'half' of the communist era state of Czechoslovakia). Effectively they were denied citizenship in what had been, until recently, their own country. *Roma* peoples are the subject of discrimination in Yugoslavia and Hungary. It is not just in eastern or central Europe that such exclusion takes place. In Britain gypsies are often denied entry to pubs and hotels simply because of the way they are perceived. In this case we can see how a particular group of people have been both formally and informally excluded from what is seen as mainstream society.

CHAPTER 1: CITIZENSHIP, INCLUSION AND EXCLUSION

Activity Box 4: Investigating exclusion

Either, (a) investigate the websites listed below – each has details on different aspects of the lives and history of the *roma* peoples in the Czech Republic:

- Helsinki Citizens Assembly – http://www.czechia.com/hcaroma/default.htm
- Roma in the Czech Republic – http://www.romove.cz/romove/situation.html
- Czech Centre – http://www.czechcenter.com/ROMA.htm
- Human Rights Watch – http://hrw.org/reports/world/czech-pubs.php

Or (b) choose a group (e.g. Kurds) that you believe are the victims of exclusion and find out as much as you can about them from a range of resources (e.g. television, the internet, newspapers/magazines).

You may wish to map their distribution across the world, continents or even countries. Provide evidence of:

- how and why they have come to be discriminated against and/or have become the victims of exclusion,
- how this has affected their settlement patterns, and
- the ways in which these people and their supporters seek to establish their identity as full citizens of a particular country (think about how persuasive their arguments are).

to participate in society. Some people may be unable to take full benefit of their rights, or they may be unaware of them. Equally it may be difficult for some to participate fully in society as a result of poverty or lack of resources. People may appear to have full citizenship rights, but may be denied economic power. Others may be denied full citizenship (Information Box 4 and Activity Box 4). They may, for example, have no control over their employer's decision to close down a factory, thereby depriving them of their right to employment and a decent income.

During 2001, riots in northern industrial towns in England such as Oldham and Bradford centred on issues of race (Figure 3). Some have argued that ethnic minority groups (in this instance people of Indian, Pakistani and Bangladeshi origin) should actively participate in the democratic political system rather than resort to violent methods to put right injustices. This view ignores the fact that some members of these ethnic groups may feel that the political system fails them. The failure of the system to protect them or to address their concerns may result in feelings of exclusion. As a consequence, they may opt to act outside the law – a violent response may be a symptom of powerlessness rather than mindlessness. Rather than merely criticising their actions it might be more appropriate to look at altering the structures so as to allow such minorities more meaningful modes of participation in society; in other words making them full citizens in reality.

Of course some groups, such as the British National Party (BNP), wish to deny ethnic minority groups any form of British citizenship, favouring instead a policy of repatriation to their 'lands of ethnic origin', as the BNP puts it (see Information Box 8, page 31). In turn, part of the reason for support of the BNP in certain areas may be attributable to a feeling of powerlessness among sections of society, feelings which all too often may become directed at easily identifiable scapegoats, such as ethnic minority groups.

Figure 3: Demonstrating for rights can take many forms and may include confrontation with the law. Photo: Guzlian Photo Library/Rob Dawson.

Summary

What all this means is that a consideration of citizenship is not just a dry and abstract subject connected to the state or to government. While there is an obvious political dimension in the narrow sense of a formal relationship between a state and its citizens, there are also much broader social and cultural issues here. Explorations of citizenship entail a consideration of the many forms of inequality which exist in the world; between different countries, within the same country and even within the same locality.

It involves some thought about who is seen to belong to a place and who is seen not to belong there. It is intrinsically geographical.

Citizenship is thus about relationships between the individual and his or her wider world. Chapters 2 and 3 discuss the state and the nation, the levels at which discussions of citizenship have tended to be placed. Chapter 4 focuses attention on those trends which might suggest that the state's pre-eminent role as a political and economic unit is drawing to a close, and the implications of that for citizenship are assessed.

CHAPTER 2

THE STATE

What is the state?

Given that citizenship is usually seen as the relationship between the individual and the state, it is important first to consider what exactly we mean by 'the state'. The best way to start is by looking at a map of the world and identifying individual countries: these are states – political-territorial entities and are regarded as the basic unit of political organisation in the world today. This was not always the case, however. Not until the Treaty of Westphalia in 1648 was the principle of sovereign states established. In addition, the state system did not spread beyond Europe until the nineteenth century, mainly as a consequence of European colonial expansion into large parts of Africa, Asia and the Americas. Before that, various forms of more localised political systems held sway.

States are not static, they are constantly changing; some disappear while new ones are created and yet others recreated (Figure 4 and Activity Box 5). With the collapse of communism in the late 1980s and early 1990s the Soviet Union disintegrated into 15 different states, the largest of which is the Russian Federation (see Figure 18). The state of Yugoslavia

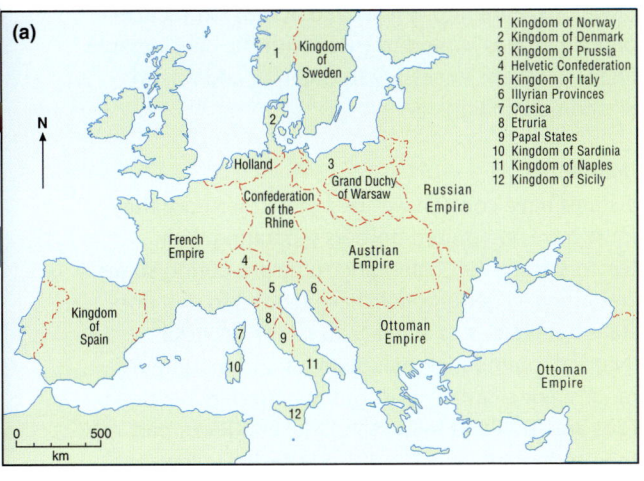

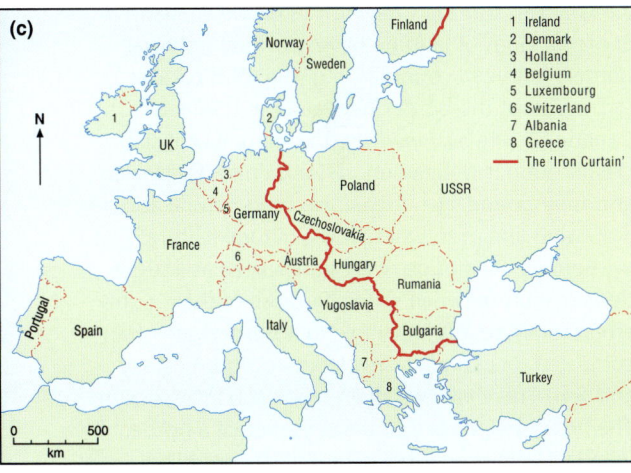

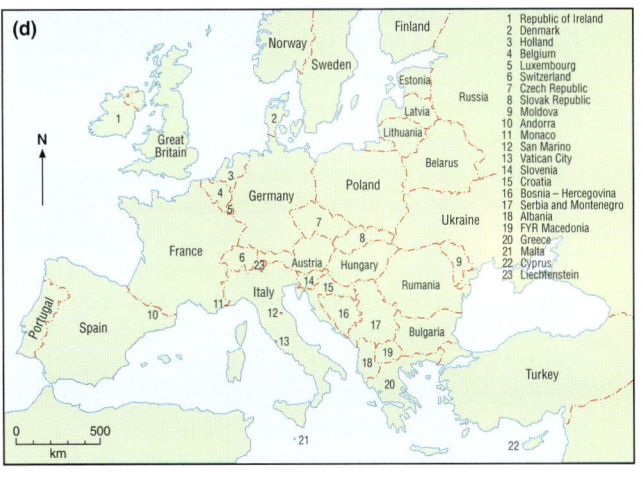

Figure 4: Changing states in Europe: (a) early nineteenth century, (b) early twentieth century, (c) mid-twentieth century, and (d) early twenty-first century. After: Heffernan, 1998.

CHAPTER 2: THE STATE

> **Activity Box 5: Changing states**
>
> Examine the maps of Europe in Figure 4 and answer the following questions:
>
> - How many states existed in Europe in each era?
> - Which states have continually existed since the early nineteenth century?
> - Which states have disappeared since that era?
> - Which states have 'disappeared' and then 're-emerged' at a later time?
>
> Choose one 'disappeared' state and one 'new' or 're-emerged' state, and use appropriate resources (history or geography textbooks, atlases, internet, etc.) to find out more about each one. Write a brief 'country profile' outlining key features and key events in each place.

also broke apart following violent clashes, and was replaced by Slovenia, Croatia, Bosnia-Herzegovina, Macedonia, and Serbia and Montenegro. In the same period, Czechoslovakia split into the separate Czech and Slovak Republics (Figure 4d).

In broad terms, a state can be defined as a set of political institutions with jurisdiction over a specified territory (the distinction between state and nation is made clear in Chapter 3, pages 26-33). A state is a spatial entity with four essential features:

- Territory
- People
- Boundaries
- Sovereignty.

Territory and people

Although the state is a political entity, it is also a *spatial* entity, comprising territory (land, air, water) over which it exercises power. It exercises that power not just over the delimited space but also over the people residing within that space or territory. States have jurisdiction over people and places. Of course the very development of a state system required a clarification of the relationship between the state and those people living within its territory. Hence, the idea of citizenship evolved alongside the evolution of the state as a territorial unit. For example, the idea of rights has been extended down from the limited concept of the rich and powerful having the right to rule others (as in monarchical and feudal systems) to a much more egalitarian idea of a liberal democracy.

One of the ideas underpinning liberal democracy is that people have the right to choose who governs them.

The power of citizens in authoritarian systems is negligible; even in many western European democracies the extent to which many people have any real control over their own lives may be extremely limited. Although, in theory, anyone can run for the office of President of the United States, in practice participation is limited to those with the considerable financial means necessary to organise and run an election campaign.

Boundaries

If states have control over designated territory they must have recognised borders separating their territory from that belonging to neighbouring states. Traditionally, geographers have tended to describe state boundaries as either 'natural' or 'artificial'. 'Natural' boundaries include rivers and mountain ranges. For example, part of the boundary between Laos and Thailand follows the Mekong River, part of the border between the US and Mexico follows the Rio Grande, and the Cheviot Hills divide Scotland from England. Lines of latitude have provided 'artificial' borders between countries, e.g. the 49th parallel forms part of the border between the USA and Canada (Figure 5). However, this natural/artificial classification is misleading. It is a human decision to impose a boundary – there is nothing natural about a dividing line even where it is a natural feature.

Borders have huge overt significance as lines of demarcation between states. Because of their highly artificial nature it is inevitable that borders are often contested, occasionally involving outright hostility and war. There are two main types of border dispute. The first concerns the actual existence of a border. In Ireland the border between British controlled Northern Ireland and the independent Irish Republic

Figure 5: 'Natural' and 'artificial' borders of the North and South Americas.

is seen by many Irish nationalists as an unwanted division imposed by Britain (Case Study 1).

A second type of border dispute has to do with the precise location of the divide between adjoining states, as exemplified by the border between Peru and Ecuador. Sections of the border between the two countries had never been adequately demarcated and a dispute over territory claimed by both countries flared up in the early 1990s, eventually being resolved in 1999 (Figure 5). In Europe, the borders between Germany and Poland and between Finland and Russia have been subject to discussion in recent years.

Of course borders are not just lines dividing territory, they are political constructions that have a significant consequence, especially for people who live in border areas. The creation and existence of formalised borders may lead to significant differences developing on either side of the line, even in places

CHAPTER 2: THE STATE

Case study 1: The partitioning of Ireland

From 1800 through to 1921 the islands of Ireland were technically part of the United Kingdom. However, British rule was contested and a series of rebellions took place in an effort to achieve independence. Following a War of Independence, the settlement of 1921 resulted in the creation of an independent Irish Free State in 1922 (now known as the Republic of Ireland), and Northern Ireland. The latter was to remain under British rule, but with its own parliament in Belfast.

The Republic of Ireland consists of 26 of the island's 32 counties, while the remaining six north-eastern counties are in Northern Ireland (Figure 6). The imposition of this division was, and remains, highly contested.

Within Northern Ireland the majority population favour remaining British and have come to be known as unionists. However, a sizeable minority population are Irish nationalists who prefer the idea of a united Ireland rather than continued British control in the North. A religious divide has compounded the unionist-nationalist tension – most unionists are Protestant and most nationalists are Catholic. Discrimination against Catholics by the unionist government eventually led to the outbreak of a period of prolonged civil unrest from the late 1960s onwards. The parliament was suspended in 1972, after which there was direct rule from London. Acts of violence were committed by paramilitary groupings on both sides, leading to the deaths of more than 3000 people over the following 30 years.

In 1998 a peace agreement was reached resulting in a power-sharing Assembly based in Belfast. However, progress has been extremely slow and the conflict has by no means been resolved (Figure 7).

Figure 6: The 32 counties of Ireland and its division into Northern Ireland and the Republic of Ireland.

Figure 7: This 'Peace for all' sculpture stands on the border between County Cavan (Republic of Ireland) and County Fermanagh (Northern Ireland). Photo: David Storey.

CHAPTER 2: THE STATE

Figure 8: The Mexico-USA border crossing point at Tijuana. Photo: David Storey.

where no major social or cultural divisions previously existed. Contrasting landscapes may evolve, including differences in settlement patterns, and people may also develop different attitudes towards the border, depending on whether they see it in a positive or negative light. While some people regard borders as means of protection from what they see as 'alien' elements outside their state, others might see them as barriers to their freedom of movement. For example, the US-Mexico border is perceived by some as protecting the USA from 'illegal' Mexicans encroaching on their territory, but it might also be read as a barrier to Mexicans, preventing them from accessing the benefits of a rich country (Figure 8).

Perhaps the most notorious border in the recent history of Europe was what became known as the Iron Curtain (see Figure 4c), an artificial divide between communist eastern Europe and capitalist western Europe, resulting from the Second World War. Germany was split into two countries, East and West, and in 1945 its former capital city, Berlin, was divided into what became known as East and West Berlin. Until 1960, many East German citizens crossed into West Germany by way of West Berlin, attracted by the relative freedom and prosperity of life in the West. To prevent this movement, in 1961 the East Germans built the Berlin

Figure 9: Germans take part in dismantling the Berlin Wall in November 1989. Photo: Gasch/Photovault.

CHAPTER 2: THE STATE

> **Information Box 5: A personal perspective on the East and West German border**
>
> In the following passage from his book, *Along the Wall and Watchtowers*, German-born writer Oliver August describes the significance of the border between West and East Germany from the perspective of his family.
>
> 'My father had been fourteen when the [Second World] war ended and the Allies drew a line across his father's tree nursery. The main house was in the Soviet zone [which became East Germany] while some of the fields were in the British zone [which formed part of West Germany]. The border literally divided the property. Aged seventeen, my father hid a suitcase on a horse-drawn cart and drove west across the border on family property, leaving his parents behind. In the following forty years he was allowed to return only twice – for a maximum of three hours each time – for their funerals' (August, 2000, p. 3).

Wall. As well as being a physical barrier to movement, the Wall also divided families between East and West (see Information Box 5), and became a powerful symbol of European disunity.

As the example of Germany demonstrates, a border which divides a formerly united country can lead to the development of significantly different political, social, cultural and economic orders on either side. The tensions and tragedies that result from such divisions are well illustrated by the quotation in Information Box 5 (see Activity Box 6). When the Wall was finally dismantled in November 1989, there were scenes of massive celebrations in Berlin (Figure 9).

Sovereignty

Sovereignty refers to the authority of a state to rule over its territory and the people within its borders: the right of the state to rule without external interference. Such interference (invasion, etc.) is seen to be a transgression of international law. Examples of transgressions include the Soviet Union's invasion of Afghanistan in the 1980s, the US invasion of Grenada in 1983, NATO air attacks on the Federal Republic of Yugoslavia in 1999 and the continued bombing of parts of Iraq by Britain and the US since the early 1990s. More recently, the legality of the 2001-02 attacks on Afghanistan and the war on Iraq in 2003 by the USA and UK have also been questioned.

For a state to be sovereign, it must be recognised as such. During the apartheid era in South Africa the white minority government created 'self-governing homelands' and 'independent republics', ostensibly devised for black residents in what was essentially a charade designed to imply the existence of political autonomy (self-government) for sections of the black population (Figure 11). These areas were not sovereign because the international community refused to recognise them as such. Only South Africa viewed (or claimed to view) them in this way. Non-white people continued to exist as second-class citizens in their own homeland (Smith, 1992). Similarly, during the so-called 'conquering' of North America by whites of European origin from the seventeenth century onwards, the indigenous 'Indian' population, as they were misleadingly referred to, did not have sovereignty. Nobody, other than the native, or first, Americans (as they are now known) themselves, recognised their claim to the land in which they lived. As a consequence, white settlers rode roughshod over the rights of the indigenous population, justifying this in the name of 'settling' the territory and bringing 'civilisation' to the 'savages'.

> **Activity Box 6: Borders**
>
> Read the poems shown in Figure 10 about borders (or other suitable poems of your own choosing). In the case of each poem consider the following questions:
>
> - Where do you think the border is and why does it exist?
> - Where do you think or imagine the character/narrator is travelling from and why?
> - Does he or she view the border in a positive or a negative light?
> - Why do they view it this way?
>
> Select one poem and write a short essay describing, in your own words, the experience communicated within it and project a future that the central character may experience. Write your essay from the migrant's perspective.

CHAPTER 2: THE STATE

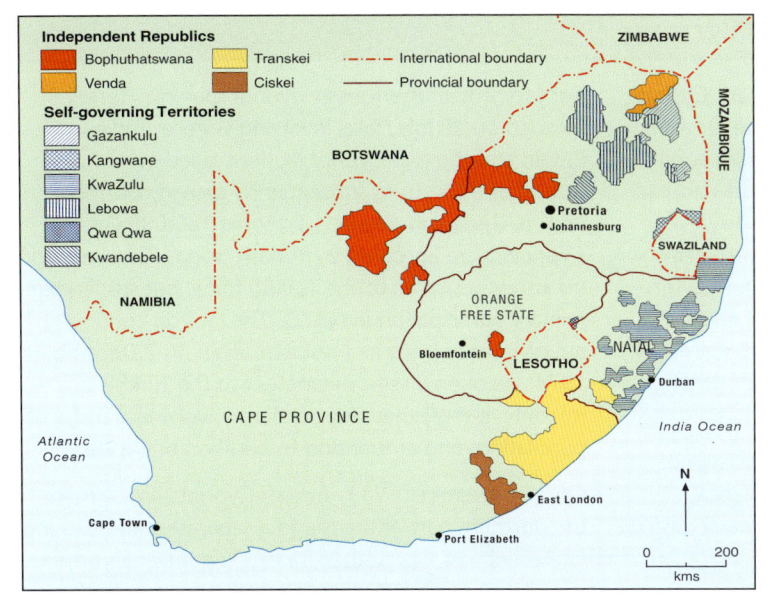

Figure 11: The South African *Bantustans*.

Instructions for Crossing the Border

By Dan Pagis (translated from the Hebrew by Stephen Mitchell)

Imaginary man, go. Here is your passport.
You are not allowed to remember.
You have to match the description:
your eyes are already blue.
Don't escape with the sparks
inside the smokestack:
you are a man, you sit in the train.
Sit comfortably.
You've got a decent coat now,
a repaired body, a new name
ready in your throat.
Go. You are not allowed to forget.

Boy with Orange: Out of Kosovo

By Lotte Kramer

A boy holding an orange in his hands
Has crossed the border in uncertainty.

He stands there, stares with marble eyes at scenes
Too desolate for him to comprehend.

Now in this globe he's clutching something safe,
A round assurance and a promised joy

No-one shall take away. He cannot smile.
Behind him are the stones of babyhood.

Soon he will find a hand, perhaps, to hold,
Or a kind face, some comfort for a while.

Figure 10: Poetic responses to the influence of borders on people's lives.
Sources: Motion, 2001; Haywood, 2002 (originally appeared in *The Phantom Lane* by Lotte Kramer. Published by Rockingham Press, 2000).

A contemporary example of the (non-) recognition of sovereignty is the so-called 'Turkish Republic of Northern Cyprus'. The 'Republic' was created following a Turkish invasion of Cyprus in 1974, in response to oppression of the minority Turkish Cypriots by the majority Greek Cypriots (see Case Study 2 and Activity Box 7).

What do states do?

The previous sections provide some idea of what states are, but we also need to consider what it is that states do. There are many different views on this. The state can be said to perform a number of functions, which are:

- regulating the economy;
- providing public goods and services such as health and transport;
- providing legal and other frameworks to guide/constrain citizens' behaviour;
- defending its territory and people against attack.

In recent years there has been a major trend in western countries towards 'rolling back the state' and privatising services that were previously 'nationalised', thereby minimising state involvement. Despite this, the influence of the state remains quite pervasive. In the United Kingdom, the state provides or subsidises education for most young people and determines the age groups for which it is compulsory; it also regulates systems such as postal delivery and the hospital care

CHAPTER 2: THE STATE

Case study 2: Cyprus

The third-largest island in the Mediterranean, Cyprus lies off the southern coast of Turkey and the western shore of Syria. For centuries its rule passed through many hands, then, in 1571, it fell to the Turks and a large Turkish colony settled on the island. However, the island's Greek population has long sought self-determination and reunion with Greece. During the First World War Britain annexed the island, and, in 1925, declared it a Crown colony. Cyprus became an independent state on 16 August 1960, with Britain, Greece and Turkey as guarantor powers.

On 15 July 1974, a military coup (led by the Cypriot National Guard) overthrew Archbishop Makarios, President since 1959. Diplomacy failed and, asserting its right to protect the Turkish Cypriot minority, Turkey invaded on 20 July. Talks involving Greece, Turkey, Britain and the two Cypriot factions failed in mid-August and the Turks subsequently gained control of 40% of the island (Figure 12). Some 180,000 Greek Cypriots were uprooted by Turkish troops. Greece made no armed response to the Turkish force, but suspended military participation in NATO. The tension continued after Makarios became President again on 7 December 1974. He offered self-government to the Turkish minority, but rejected any solution 'involving transfer of populations and amounting to partition of Cyprus'.

On 15 November 1983, Turkish Cypriots proclaimed the northern part of the island a separate state, naming it the 'Turkish Republic of Northern Cyprus'. In the same month the UN Security Council declared this action legally invalid and called for Turkey's withdrawal. No country except Turkey has recognised this illegal entity as a 'state'.

Cyprus has a good chance of joining the European Union. It has met all the economic standards, but the continued strife between the Greek and Turkish Cypriots threatens the island's potential membership.

Source: Infoplease website.

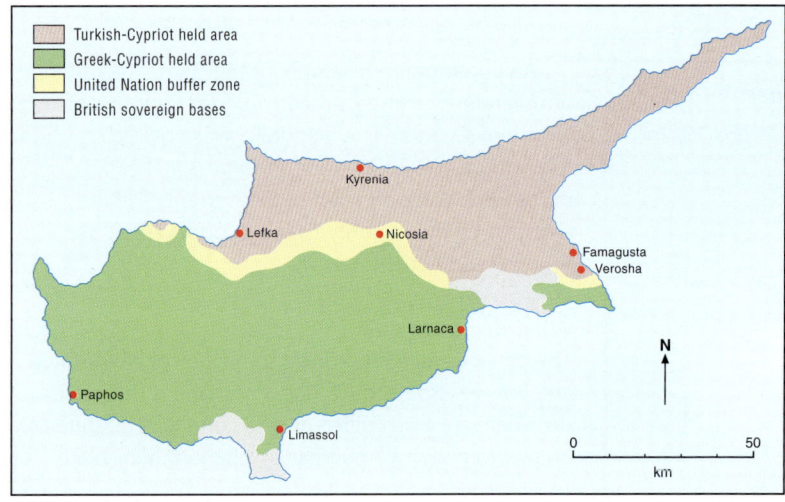

Figure 12: The division of Cyprus.

Activity Box 7: Divided space

In groups, consider the case of Cyprus (see Case Study 2 above) or another politically divided territory such as Kashmir or Israel/Palestine and discuss the following questions:

- Why do these divisions occur?
- Are these divisions real or imagined?
- Is the construction of borders between divided groups a realistic solution to these problems?
- What other solutions might be possible?
- How might these be brought about?

Each group should feed their responses back to the whole class. In your feedback you should outline issues where there was agreement within the group and those where there was disagreement.

which we receive; and it employs members of the police and fire services. It also exercises either direct control (provides the service) or indirect control (through the regulatory framework) over many of the services we receive (e.g. television, telecommunications, energy supplies) (Activity Box 8).

Activity Box 8: State regulation

Think about the things you do on a regular basis (going to school or college, shopping, participating in sports, etc.). Consider the various ways in which the state regulates what you do, and how and when you can do it.

- Draw up a list of state regulations which directly affect you.
- Discuss these in class and then on your own consider and record what would happen if these rules did not exist.
- Would it be a good idea to cancel state rules and regulations? What might be the potential problems? Give reasons for your answers.

As well as defining the way states function, we can explore the *meaning* of 'state' at a deeper level. There are different viewpoints about the nature of the state, and the way a state should behave. These fall into two broad, and opposing, categories – pluralist and Marxist.

The pluralist views the state as a neutral arbiter that resolves issues arising between individuals and interest groups within its borders. Pluralists see the state as an impartial entity refereeing between competing sets of vested interests. In this way of thinking, the state is seen to mediate between the needs of individuals and groups within society and, to balance those against collective needs such as security, health care, defence and so forth.

Pluralist views of the neutral state are criticised by proponents of a Marxist theory of the state. Marxists argue that in a capitalist society the state functions in a way which preserves the existing socio-economic order, thereby preserving capitalism. It is argued that the function of capitalism is the accumulation of profit, and that the state functions in a way that preserves the interests of the capitalist class (e.g. the owners of industry, large businesses), while working-class people are exploited through selling their labour. Profit accrues through maximising the difference between the value of output and the wages paid to workers. The state, it is argued, serves to sustain and regulate this system and helps to reproduce this capitalist ideology or way of viewing the world (Taylor and Flint, 2000).

State control is bolstered through various systems. The educational system is one example. Ideas of loyalty to the state and the desirability of obeying the laws of the land and other civic values are inculcated into people from a very young age. For example, in the USA many schools require their students to swear the Oath of Allegiance every morning (see Information Box 3, page 9). Some see this as a subtle type of indoctrination, which is further bolstered by the various media (e.g. newspapers, television). The media are generally perceived as presenting a 'commonsense' view of the world, thus certain viewpoints are considered to be off-limits, and are characterised as extreme and, hence, not practical, or appropriate. At the time of writing the USA and the UK are at war with Iraq. Much political and media attention is devoted to portraying the alleged threat the Iraqi regime presents to the rest of the world. In most media discussion little reference is made to the fact that it is the USA which is attacking another country. In much reporting, US and UK official briefings tend to be presented as facts while Iraqi statements tend to be regarded as propaganda which needs to be treated with caution. In the recent Afghanistan conflict, US claims of 'successful' bombing missions were widely reported as such in the west. However, little attention was paid to the number of civilian casualties, which are now estimated to be well in excess of the numbers killed in the attacks in New York and Washington on 11 September 2001.

This situation is referred to as hegemony – a concept developed by the Italian Marxist, Antonio Gramsci (1891-1937). Gramsci referred to 'the "spontaneous" consent given by the great masses of the population to the general direction imposed on social life by the dominant fundamental group' (1971, p. 12). This is very similar to what Michel Foucault means when he says 'each society has its regime of truth, its "general politics" of truth: that is, the types of discourse which it accepts and makes function as true' (1980, p. 131). Governments are very selective in the actions they take; for example, in the late 1990s western European states were particularly keen to punish Yugoslavia for its harsh actions against Albanian Kosovars. Likewise, the USA attacked Afghanistan for sheltering Osama Bin Laden who was blamed for the attacks on the World Trade Centre and the Pentagon on 11 September 2001. At the same time, the US and UK are forging closer

CHAPTER 2: THE STATE

Figure 13: The Caucasus region, including Chechnya.

> **Activity Box 9: Reliable sources**
>
> Consider the various ways in which you receive information about local, national and international affairs (see e.g. Figure 14). Either over seven or fourteen days look at a number/cross-section of sources in detail.
>
> - What are your sources?
> - How accurate or reliable do you think these sources are, and why?
> - How do you form opinions?
> - How do you decide what to believe?
> - How do you verify the accuracy of what you are told?

links with Russia, despite that country's activities in Chechnya, where there has been an attempt to break free of Moscow's control (Figure 13). Chechen separatists have waged war against Russian control and one group recently took people hostage in a Moscow theatre (Figure 14). This attempt by Chechnya to gain independence has been met with fierce resistance from Moscow.

The subtle ideological control of its people by a state is bolstered through systems of social welfare, provision of unemployment assistance, pensions and the like. Through these devices, the state manages to deflect much opposition to its policies. It can be argued that the risk of industrial unrest, protest marches and other indicators of opposition or resistance are minimised through the provision

CHAPTER 2: THE STATE

Chechen gunmen storm Moscow theatre

Nick Paton Walsh in Moscow and Jonathan Steele writing in The Guardian (24 October 2002)

Up to 700 people were taken hostage last night when a group of heavily armed men and women stormed a packed Moscow theatre during a musical, firing shots in the air.

Russian police said that the gang of up to 50 assailants demanded the immediate withdrawal of Russian troops from Chechnya, and said they were prepared to die for their cause.

They burst on to the stage during the second act of a hit musical, firing shots into the ceiling. They ordered the cast off stage, then told all children to leave the theatre. There were reports that Muslims had also been allowed to leave.

Members of the audience said that the gunmen had land mines strapped to their bodies and had drilled holes in the theatre structure and filled them with explosives.

The gunmen – who are said to be led by a nephew of Chechen warlord Arbi Barayev – burst into the theatre in the south-east of the Russian capital after arriving in a white minibus at 22.15pm local time (19.15pm BST).

Chechen gunmen hold 700 hostage in Moscow theatre

Fred Weir in Moscow writing in The Independent (24 October 2002)

Up to 40 heavily-armed masked men and women, believed to be Chechen rebels, are threatening to blow up 700 hostages in a Moscow theatre unless Russian troops withdraw immediately from the breakaway republic.

The gunmen stormed the former House of Culture on Melnikova Street in south-east Moscow, during a performance of a popular musical Nord-Ost. The attackers fired shots and ordered everyone to be seated.

Witnesses said they released about 20 children and Muslims, before planting booby traps around the 1,000-seat theatre's doors and windows. One witness said the guerrillas strapped explosives to the internal supporting columns of the theatre and threatened to blow up the building if police stormed it.

There were reports that at least one Russian police officer had been killed and that an explosion was heard in the vicinity of the theatre early this morning.

Hostages' calls to families create a tapestry of fear

(The Times, Friday 25 October 2002)

For just a few minutes yesterday, Karina managed to contact her sister trapped inside the Moscow theatre. Gripped by fear and speaking softly to avoid attracting attention, Alona, 32, had described how the armed Chechen rebels had stormed the auditorium.

Moscow: 'They die at dawn'

(The Sun, Saturday 26 October 2002)

Crack Russian troops launched a daring raid to save the theatre hostages early today. Armed soldiers stormed the theatre where 700 have been held hostage. The terrorists issued a warning. The 700 terrified hostages in a Moscow theatre were told by their captors they would die at dawn today.

Figure 14: Newspaper reports help indicate how some groups use extreme and threatening behaviour in order to obtain international recognition for their fight for independence.

of sufficient incentives to 'buy-off' the otherwise disaffected. In other words, while some groups may be unhappy with such problems as unemployment, short-term contracts, or relatively low wages, the state is seen to provide a 'safety net', and this helps to reduce and prevent serious unrest. It can be argued that welfare programmes and other attempts by the state to provide some form of assistance to the less well off might be seen as efforts at maintaining political stability rather more than genuine attempts to alleviate poverty.

Summary

The state is something more than a provider of services. It is a political-geographical entity wielding power over people in a defined geographic space. Relating all this back to the idea of citizenship, we can say that the state, through its various activities, grants certain rights to its citizens in return for their loyalty to it. This loyalty is ensured through various means, some of which have been discussed in this chapter. Loyalty to the state is more likely to occur if there is also loyalty to the nation. The concepts of nation, national identity and nationalism are the focus of the next chapter.

CHAPTER 3

THE NATION

What is a nation?

As we have seen in Chapter 2, a state can be defined as a set of political institutions which operate within a clearly defined territory. A nation, on the other hand, may not have a defined territory, and people of the same nationality may occupy more than one state. Nation is the term we give to a group of people, a national community, who share certain things in common – descent, history, culture, language, and so on.

There is really no such thing as a nation-state, that is, a territory occupied and run by one national group. More usually, states are occupied by people from many national groups, some of whom may be content to regard themselves as belonging to that state, but others may not – as we shall see later in this chapter. So, for example, there are people who 'feel' French, Portuguese, Malaysian, and so on, and their national communities are linked through their occupation of, or association with, a particular

Information Box 6: The Basques

The 'Basque country' in northern Spain and south-west France is the name of a territory occupied by a people who see themselves as part of the historic Basque nation (Figure 15). For many of them their Basque-ness takes precedence over their Spanish-ness or French-ness. Their sense of Basque identity is heightened by the regional use of the Basque language, which is very distinctive and unrelated to either French or Spanish.

From the 1930s to the early 1970s, the Spanish dictator General Franco banned the use of the Basque language. In an effort to preserve Spanish unity and to bolster his own personal support Franco also outlawed expressions of regional distinctiveness.

In 1937, collusion between Spanish Fascists and Hitler's Nazi regime led to the bombing of the unprotected Basque town of Guernica by the German *Luftwaffe* (Air Force). (The artist Pablo Picasso immortalised this event in a painting he produced in response to the bombing, entitled *Guernica* (see Treasures of the World website, page 48)). This violent repression had quite the opposite effect to that which Franco desired; it led to the emergence of a strong Basque independence movement and to the formation of a paramilitary grouping known as ETA. Since the death of Franco, the Basque country has achieved a level of autonomy within Spain's federal system. However, this falls short of the total independence still demanded by ETA and others (Figure 16). ETA continues to carry out violent attacks on politicians, police and other targets in pursuit of their political goals (see Collins, 1990; Kurlansky, 2001).

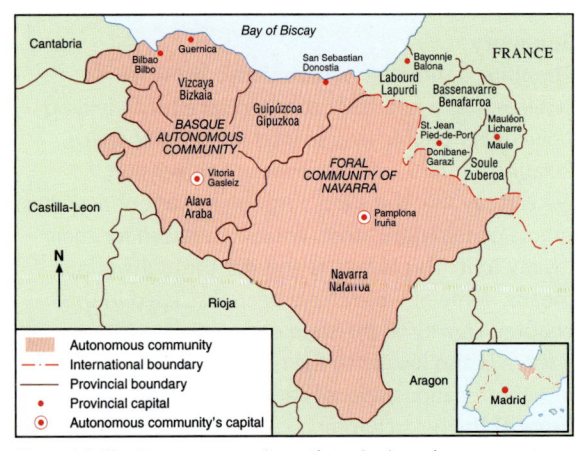

Figure 15: The Basque country in northern Spain and south-west France.

Figure 16: This call for Basque independence, has been posted in the city of Donostia/San Sebastian. Photo: David Storey.

territory. Thus, French people who live in different parts of the world may still regard France as their 'homeland'.

In the same way, people of different nationalities who live in one state may regard themselves as belonging to that state. In the United Kingdom, for example, as well as the Welsh, Scottish, English and (Northern) Irish, there are numerous other national communities such as Indian, Bangladeshi, Jamaican, and Chinese, many of whose members regard themselves as being British, but some who do not.

In the case of France, 7% of the population are not considered to be French. Included in this group are immigrants from Algeria, Morocco and Tunisia in northern Africa. Members of these ethno-national minorities are often victims of racism in France. There are also those, such as Bretons (in north-west France), and Basques (in the south-west of France) who do not see themselves primarily as French, despite their long association with the state of France. What both these groups have is a strong feeling of belonging to a regional/national community, but one which has limited 'official' recognition.

In some sense, then, a nation is a mental construct – an idea to do with identity which is usually linked to a particular territory. One way of thinking of this national identity is as a glue which binds people together. It is something which most people feel strongly about, and this is why states can use the idea of nationality to mobilise people and encourage them to take action, such as defending their territory against aggression from 'outsiders'. Viewed in this way, citizenship might be seen as a means of inculcating loyalty to the state through support for the nation. In return for their allegiance to the nation/state, citizens are granted protection, security, etc., in a way which helps to bind nation and state together.

There are many examples around the world of this sense of nationhood being harnessed by the state to defend its territory against people of different nations and/or states. A recent example in Europe is the case of Kosovo in former Yugoslavia.

Here, one group (the Serbs) were encouraged by their leader, Slobodan Milosevic, to fight against the Albanians on the grounds that the whole territory of Kosovo belonged to the Serb nation. The result was a terrible loss of life among Albanians who also felt that they had a just claim to territory and were seeking autonomy for Kosovo (see Storey, 2002).

David Miller, in his book *On Nationality*, describes the relationship between nation and state as follows:

> '"Nation" must refer to a community of people with an *aspiration* to be politically self-determining, and state must refer to the set of political institutions that they may aspire to possess for themselves' (1997, p. 19, emphasis in original).

When a nation has a strong 'aspiration to be politically self-determining', it is possible that it will wish to secede from the state in order to achieve full nationhood and independence. It is just such an aspiration that drives the Basque separatist movement in northern Spain (see Information Box 6).

Likewise, the Kurds live in a number of different states. The present 'heartland' of the Kurdish nation (called Kurdistan) lies in the border territory of at least four states – Turkey, Syria, Iraq and Iran (Figure 17). In these countries, Kurds have been subject to persecution in some cases and are mostly treated as second-class citizens. What the Kurds desire is their own independent state.

Figure 17: The extent of the Kurdistan 'heartland'.

CHAPTER 3: THE NATION

Figure 18: The independent republics that made up the former Soviet Union.

In the United Kingdom, both Scotland and Wales were given a measure of autonomy (self-governance) in the 1990s (Scotland now has its own Parliament, and Wales has a National Assembly), but the Scottish and Welsh nationalist parties continue to campaign for full independence for their 'nations'.

The former Soviet Union is yet another example of a multi-nation state. When it was broken up in 1991, a number of independent Republics were formed (Figure 18). Many of these, e.g. Estonia, Latvia, Lithuania, were former states, and each had a strong national identity of its own. However, the population of these states is by no means homogenous, that is, not all the inhabitants of Estonia are Estonian, many are Russian or of other nationalities. Indeed, full citizenship of Estonia is denied to those of Russian origin (Unwin, 1999) (Activity Box 10).

Let us return to the idea of nations as mental constructs, or as Benedict Anderson (1991) calls them, 'imagined communities'. Anderson argues that nations are 'imagined' because no one member of a nation will ever know or meet all of the other members, yet she/he knows that they exist. However, despite its constructed nature, the idea of the nation has a very real resonance for many people. For example, 'national identity' is always very clearly expressed at international sports events such as the football World Cup. A common ritual at these events is to sing national anthems, wave national flags and generally support the 'national' team. The national bond is very strong on such occasions, and they can become quite emotional.

Activity Box 10: Conflict in multi-nation states

Study Figure 18 and choose two states that became independent republics after the break up of the Soviet Union in 1991. Using the internet and other references sources find out what you can about the different national groups that live in these independent republics.

- Is there any evidence of conflict between the different national groups?
- If so, what form does this take, and how is it being dealt with?
- Explain in your own words why you think such conflict exists.

Activity Box 11: Defining your 'nationality'

Consider and record your responses to the following questions:

- What nationality do you think of yourself as having, and why?
- Are all or most of your friends of the same nationality, and if not, which nationalities are they?
- Are there times when your feelings of national identity are stronger than others? If so, when, and why do you think this is?

Discuss your responses with other students (and your family if you wish) and compare your answers. Are there any similarities or differences? Is it easy to explain them?

Nationalism and national diversity

Anthony D. Smith defines nationalism as 'an ideological movement for attaining and maintaining autonomy, unity and identity on behalf of a population' (Smith, 1991, p. 73). Such movements may be peaceful but others may be violent, especially where there is a strong sense of grievance against the state. Such militant nationalism has led to thousands of deaths, injuries and much devastation all around the world (as in the case of Northern Ireland and Chechnya, for example). It is as a consequence of this that nationalism is regarded by many as regressive and dangerous.

Whatever we may feel about nationalism, it clearly involves very deep-seated feelings which lead people to go to great lengths to assert their identity or defend their right to express it (go to Activity Box 11.)

A sense of national identity may be expressed – and reinforced – in positive, subtle, and even banal ways (Billig, 1995). For example, the flying of flags over public buildings and during ceremonies (Figure 19), the existence of national television stations, national newspapers and so forth all serve to express the idea of the nation as a concrete entity. Similarly, dress, food, language, daily habits, festivals and a variety of other things may be associated with nationality (Activity Box 12) – as well as people's names (e.g. Kelly and Murphy are assumed to be Irish names, Perez Spanish, Hoffman German, and so on). These kinds of differences serve to reinforce our sense of a world comprised of nations. In the words of Michael Billig: 'the world of nations is the everyday world' (1995, p. 6).

Of particular interest to geographers is the way in which particular images of place may directly or indirectly shape a national consciousness. In the case of England, for example, images of idyllic rural life are often used to promote the 'essence' of England, and by association, of 'being English' (Information Box 7). These images and perceptions often bolster a very narrow conception of national identity. This may serve to engender feelings of exclusion amongst those seen not to conform to these ideas. Black people, in

Figure 19: The flying of flags is just one of way demonstrating a national identity. On 27 September 2002 the Democratic Republic of Timor-Leste became the 191st member of the United Nations. Photo: Mark Garten/United Nations.

CHAPTER 3: THE NATION

> **Activity Box 12: 'Local' national identity**
>
> Following from Activity Box 12, think further about the ways you and your friends (and family) conceive of your national identity.
>
> - In what ways does national identity impinge on every-day aspects of your life?
> - Walk around the streets in the centre of your nearest city/large town and list the various signs/features in the urban landscape which signify national identity.
> - What do these features tell you about the 'embeddedness' of national identity?

particular, may feel isolated or vulnerable in rural England, whether living or just visiting such areas (Kinsman, 1995).

There are many positive aspects to national diversity, and the world is a richer place for it. However, problems can occur when questions arise as to who does and does not belong to a nation, or when one group considers itself to be superior to another (national chauvinism). The sense of superiority may be created and/or reinforced in a variety of ways; for example, in Britain, by the singing of 'Rule Britannia' at public events. This song evokes images of what some see as Britain's glorious imperial past, of which its people should be proud, and it can bring tears to the eyes, even of those who would not otherwise see themselves as nationalistic. However, for other people (such as some of those who grew up in British colonies or former colonies) the song may be regarded as more than rousing and harmless music; for them it may symbolise Britain's sense of superiority over other nations and states – including those from which they or their ancestors originated.

In Britain, many non-white communities feel excluded, discriminated against and even afraid, and such feelings are reinforced by the activities of far-right groups such as the National Front and the British National Party (BNP) (see Information Box 8). The phrase 'there ain't no black in the Union Jack' sums up both the sense of exclusion among many black people, and also the wish of others to exclude them from 'the British nation'. There are those in Britain who consider that anyone who is black cannot be 'really' British, despite the fact they were born in Britain (Gilroy, 1987). This line of racist thinking is both simplistic and dangerous. It is simplistic because it ignores the complex nature of population movements and the cultural influences that modify the ways in which people live. It is dangerous because it breeds negative rather than positive feelings about difference (Activity Box 14).

Britain has long been a multi-nation state, comprising three distinctive national communities with widely different cultures and heritage – the Welsh, Scottish and English. Over time, various aspects of each 'nation' have been 'diluted' due to the powerful influence of one of the others. In the case of Wales, for instance, the demise of the Welsh language is sometimes attributed to the pervasive

> **Information Box 7: The rural idyll**
>
> Below is an excerpt from *In Search of England* by the travel writer H.V. Morton. Originally written in the 1920s, it demonstrates the author's vision of an England to which he was longing to return at a time when he was out of the country.
>
> 'There rose up in my mind the picture of a village street at dusk with a smell of wood smoke lying in the still air and, here and there, little red blinds shining in the dusk underneath the thatch. I remembered how the church bells ring at home, and how, at that time of year, the sun leaves a dull red bar low down in the west, and against it the elms grow blacker minute by minute. Then the bats start to flicker like little bits of burnt paper and you hear the slow jingle of a team coming home from the fields ... This village that symbolizes England sleeps in the sub-consciousness of many a townsman [sic]. A little London factory hand whom I met during the [First World] war confessed to me ... that he visualised the England he was fighting for ... as not London, not his own street, but as Epping Forest, the green place where he had spent Bank Holidays ... The village and the English countryside are the germs of all we are and all we have become: our manufacturing cities belong to the last century and a half: our villages stand with their roots in the Heptarchy' (Morton, 2000, pp. 1–2).

Activity Box 13: Rural idylls and personal identity

Think about the different ways in which rural areas in Britain are portrayed (Information Box 7 provides one example and Ingrid Pollard supplies a contrasting view – see Websites). Examine a range of 'texts' that deal with or are set in the British countryside, such as:

- paintings (e.g. by John Constable, J.M.W. Turner), poetry (e.g. Dylan Thomas's *Under Milk Wood*, William Wordsworth's *Lines*), magazines (e.g. *Country Living, Rock+Run, High*);
- television programmes (e.g. *Two Thousand Acres of Sky, The League of Gentleman, Take the High Road, Down to Earth*);
- radio programmes (e.g. *The Archers*);
- films (e.g. *Ryan's Daughter, Local Hero*) and books (e.g. *The Englishman Who Went Up a Hill but Came Down a Mountain*, Monger, 1995; *On the Black Hill*, Chatwin, 1998);
- various countryside organisations' websites (e.g. The Countryside Alliance, Ramblers' Association, Campaign for the Protection of Rural Wales, The Council for the Protection of Rural England, the Association for the Protection of Rural Scotland).

For each 'text', try to answer the following questions:

- How realistic do you think these representations are?
- What is included or excluded from these images?
- How do these representations link to ideas of national identity?

For more detailed consideration on perceptions of the countryside, see Richard Yarwood's *Changing Geography: Countryside Conflicts*, 2002.

---

influence of an English-speaking culture. The result is that the Welsh language remains a minority language in Wales, spoken by only about 20% of the Welsh population (Aitchison and Carter, 1994). However, the teaching of Welsh in schools is now on the increase.

So what does 'being Indian' or 'being British' mean? As discussed earlier, it is about a sense of identity, but there are many constituents to that identity and it can change over time as different elements are added to the mix. This is why it is misguided to assume that there is something static about the characteristics of any particular nation. National cultures are not static; they borrow heavily from each other and are constantly evolving.

Hybrid national identities

If you did the activities for Activity Box 11 you will know how difficult it is to slot people into one category of nationality (e.g. British, Irish, Jamaican). What may be more productive and more realistic is to think in terms of hybrid identities. A person may feel they are part West Indian or Pakistani and part British, part African and part American and so on.

Information Box 8: British National Party policy

Despite the obvious difficulties in identifying any idea of national 'purity', parties such as the British National Party (BNP) still seem to believe in this idea, as shown by the following quote from the BNP website:

'On current demographic trends, we, the native British people, will be an ethnic minority in our own country within sixty years. To ensure that this does not happen, and that the British people retain their homeland and identity, we call for an immediate halt to all further immigration, the immediate deportation of criminal and illegal immigrants, and the introduction of a system of voluntary resettlement whereby those immigrants who are legally here will be afforded the opportunity to return to their lands of ethnic origin assisted by generous financial incentives both for individuals and for the countries in question. We will abolish the 'positive discrimination' schemes that have made white Britons second-class citizens. We will also clamp down on the flood of "asylum seekers", all of whom are either bogus or can find refuge much nearer their home countries' (BNP, 2003).

Activity Box 14: Politics and 'nationality'

Consider the BNP statement in Information Box 8. Does it contravene the United Nations Declaration of Human Rights (see Information Box 1, page 7)? If so, in what ways does it do so?

Using the internet and/or printed documents from your local library, find out about the policies of other UK political parties towards national identity, citizenship and immigration.
- Are their polices very clear and explicit or are they vague?

Compare the policies and note the main differences between them (if any).
- Are their policies similar to or different from those of the BNP? In what ways?

As a whole class, debate the assertion that 'Nationality is not important in the twenty-first century'.

This idea of a hybrid national identity is most apparent in the context of international sport. In many sports there has been controversy surrounding players who choose to represent countries other than the one in which they were born. Examples include the boxer Lennox Lewis and the tennis player Greg Rusedski, both of whom were born in Canada but compete for Britain. In football, the Republic of Ireland has included a sizeable number of British-born players in their international teams over the past 15 years (Holmes and Storey, forthcoming). Similarly, at the 1998 World Cup, a number of members of the Jamaican squad were born in England, but had opted to play for the country of their parents' birth (Activity Box 15). As Table 3 shows, during World Cup 2002 a number of players represented a country other than the one in which they were born.

What these examples tell us is that for many people their being or sense of self has been shaped by more than one culture. For them being forced to identify with 'one or the other' may be inappropriate and even offensive.

In many parts of Britain, people regularly celebrate their 'homeland' national identity, for example, the Notting Hill Carnival held in London, and Chinese New Year celebrations are held in many big cities. Likewise, the sizeable Irish communities in the United States hold large St Patrick's Day parades in New York (Figure 20), Boston and other cities on 17 March, while Scottish communities in Canada and elsewhere hold special suppers on Burns Night, 25 January. Within the hybridisation of identities and places, groups of people still wish to associate with a particular distinct characteristic of their own 'nationality'.

Summary

While it should be clear that the nation is more of a mental construct than a concrete reality, this is not to suggest that it is 'unreal'. Nations exist in the sense that various groups of people perceive themselves as

---

Activity Box 15: World Cup players

At the 1998 football World Cup, Jamaica selected a number of British-born players of West Indian parentage to represent them. One of these players, Robbie Earle, was born in Stoke-on-Trent (England) and played for his local club Port Vale and later Wimbledon (a London club). Here is Robbie Earle, reflecting on the situation:

'I have always believed that the English-based players in the Jamaican squad are representing not only the population of the island [of Jamaica] but all those exiled Jamaicans in Britain and across the world. Second-generation Jamaicans playing in the World Cup finals are a fitting tribute to those who sailed to Britain ... And this has been brought home to me by the numbers of well-wishing letters I have received on my return from trips away. Many exiled Jamaicans are realising their dreams through the team, and that means a lot to us' (Earle and Davies, 1998, p. 108).

- Draw up a list of sportspeople who have played for a country other than their country of birth (in addition to the footballers listed in Table 3).
- What does this suggest to you about the relationship between national identity and sport?
- Do you think these people were motivated by career ambitions or by cultural affinity?

CHAPTER 3: THE NATION

Table 3: Country of birth and international affiliation for selected football players, World Cup 2002.

Player	Country of birth	Country played for in World Cup 2002
Ümit Davala	Germany	Turkey
Owen Hargreaves	Canada	England
Matt Holland	England	Republic of Ireland
Muzzy Izzet	England	Turkey
Miroslav Klose	Poland	Germany
Oliver Neuville	Switzerland	Germany
Emmanuel Olisadebe	Nigeria	Poland
Efe Sodje	England	Nigeria
Branko Strupar	Croatia	Belgium
Patrick Vieira	Senegal	France

possessing shared histories, values and cultural legacies. The nation is a construct which on the one hand may allow some people to feel part of a wider entity but, on the other hand, may leave some people feeling excluded from full membership of that entity.

While the existence of different nationalities may well lead to a richer cultural mix, nationalism can also be an exclusivist ideology, serving to disempower certain categories of people. In the next chapter we turn attention to other influences on the nation and state.

Figure 20: Celebrations of national identity: (a) St Patrick's Day Parade through the streets of New York. Photo: James Casserly 973 763 1409, and (b) Notting Hill Carnival, London. Photo: © Timothy N. Holt, 2002.

CHAPTER 4

GLOBALISATION, STATE AND NATION

As discussed in Chapter 3, the well-being or continued existence of some states may be threatened from within their boundaries. States may also be threatened from the outside, by extra-national or global processes. These processes are usually grouped together under the umbrella term 'globalisation'.

Globalisation refers to the flow of, and interaction between, people, capital, goods, labour, ideas, fashion, information, and so on, around the globe. Information about, and direct communication with other people and places is readily and constantly available to most of us (at least those of us living in the 'developed' world). This is why we think of the world as 'shrinking', or getting smaller. Advances in telecommunications and information technology, and cheaper and quicker air travel, for instance, combine to bring previously 'remote' places closer to us.

Some geographers have likened globalisation to time-space compression: in an increasingly interdependent and interlinked world, fashion and other trends diffuse quickly around the globe and permeate political and cultural boundaries quickly and easily (Harvey, 1989; Massey, 1994). Distance between places ceases to be a barrier to communication and diffusion. It could be argued that territory and borders have diminished in significance, and that the concepts of nation and state are increasingly irrelevant as we move towards an ever more cosmopolitan world. In such a world, questions of nationality and place become less and less important. By implication, the relationship between an individual and the nation/state is being weakened. Perhaps the idea of the citizen as previously understood is becoming obsolete. In effect, we may need to consider ourselves global citizens, rather than citizens of states or nations.

Dimensions of globalisation

For the sake of convenience, we can look at globalisation in terms of different dimensions, or processes – economic, political, socio-cultural and environmental. However, it is important to bear in mind that in reality these are interlinked. So, for example, we need to consider the social, cultural or environmental impacts of economic processes, or the links between environment and politics, and so on.

The economic dimension

It is in terms of economics that we see some of the most obvious signs of globalisation. For example, there has been considerable growth in the extent of international trade and of foreign investment. The rise of multi-national or trans-national corporations (MNCs or TNCs) is in itself evidence of increasing cross-border economic activity. These are large companies which own or control production facilities in more than one country, and can open and close down branch plants with relative ease (Dicken, 1998). A major MNC, such as the US-based Microsoft, will

---

Information Box 9: The power of multi-national corporations

- Of the 100 largest economies in the world, 51 are multi-national corporations (MNCs) and only 49 are countries (based on a comparison of corporate sales and the gross domestic product (GDP) of countries).
- The Top 200 MNCs' sales are growing at a faster rate than overall global economic activity.
- The Top 200 corporations' combined sales are bigger than the combined economies of all countries excluding the biggest ten.
- The combined sales of the Top 200 amount to 18 times the combined annual income of the 1.2 billion people (24% of the total world population) living in 'severe' poverty.
- While the sales of the Top 200 are the equivalent of 27.5% of world economic activity, they employ only 0.78% of the world's work force.
- Between 1983 and 1999, the profits of the Top 200 firms grew by 362.4%, while the number of people they employed grew by only 14.4%.

Source: *No Logo – http://www.nologo.org*

have branch offices and plants in many countries worldwide. These large companies account for a sizeable amount of world trade, up to one-quarter according to some estimates (Held *et al.*, 1999).

Many of the world's largest corporations are wealthier than some less economically-developed countries (LEDCs) (Information Box 9), and this makes them extremely powerful. They employ thousands of people world-wide, and they can help raise economic output in the countries in which they operate. For these reasons, MNCs are generally welcomed by governments, despite widespread criticism of their methods of operation. For example, such companies can and do exploit cheaper labour, and the more lax labour laws or environmental regulations in certain countries. They may also be able to force governments to grant certain concessions, such as tax 'holidays', low rates of corporate taxation, and start-up grants. In addition, they could de-stabilise local economies if they withdrew production plants in order to set up in places which offer more favourable conditions.

Many large companies have adopted a strategy of contracting out the actual manufacture of their products to sub-contracting firms, most of which operate in LEDCs employing extremely cheap labour. Many of these companies operate in Export Processing Zones. These are exempt from national legislation and are notionally intended to generate employment through the production of products for markets in MEDCs. As illustrated in Case Study 3, job security is not guaranteed and employment conditions in the factories in these Zones may be very poor (Activity Box 16).

Another example of how state borders may be 'dissolved' for the purposes of enabling economic activity is through the formation of regional trading blocs. These comprise a group of (usually) neighbouring countries that have a formal economic (and sometimes political) alliance. The European Union is an example of such a bloc. It began as the European Economic Community in the late 1950s (members were Belgium, France, Italy, Luxembourg, Netherlands and West Germany), but has evolved into a more overtly political arrangement with a current membership of 15 countries and a waiting list of applicants or candidate countries (Figure 24). In some respects the EU is beginning to take on the nature of a 'super-state', as power is taken away from the constituent members and vested in Brussels (i.e. the administrative and governing centre of the EU). Similar trading blocs and agreements exist in other parts of the world, such as the Association of South East Asian Nations (ASEAN), and the North American Free Trade Agreement (NAFTA). However, none is as well developed as the EU.

In the case of the EU, the aim is for free movement of goods, information, capital and labour within the 'community' of member states. However, this move towards equal internal access is accompanied by an increasing 'hardening' of the external borders of the EU. For example, entry of particular goods, such as certain foodstuffs, may be met with prohibitive common tariffs. There are also barriers to immigration into EU countries. This is why the notion of 'Fortress Europe' has developed; the external 'perimeter fence'

Activity Box 16: Multi-national corporations

Working in groups, use a variety of sources (e.g. websites, the company's Annual Report, newspaper/magazine articles, books) to undertake an investigation of one MNC. Your aim is to learn about the global scale of its operations, and about the methods it uses to create and maintain its wealth.

- Start by identifying one of its products and tracing the various stages of production from the origin of the raw materials used, through processing those materials to final production and sale of the end product.
- On a world map, plot the locations involved in each stage (including the location of the company's offices) (see e.g. Figure 22).
- Try to obtain information about the numbers of employees, and conditions of employment, in the various parts/stages of the business.
- Identify which aspects of the working conditions in the factories could be said to contravene the *Universal Declaration of Human Rights* (see Information Box 1 on page 7), which relate to employment, and Information Boxes 9 and 11 and Case Study 3.

Give a presentation of your findings to the rest of your class (using ICT wherever possible). After all presentations have been given, use the world maps produced by each group as the focus for a display on 'The global reach of MNCs'.

CHAPTER 4: GLOBALISATION, STATE AND NATION

Case Study 3: Game, set and match!

One MNC that has shifted the manufacture of one of its products is Dunlop Slazenger. The company, supplier of tennis balls for Wimbledon since 1902, has now moved tennis ball production from Barnsley, England, to Bataan in the Philippines.

'During Wimbledon fortnight 560 players use 48,000 tennis balls [but] few of those watching give more than a passing thought to the furry greenish yellow objects hurtling to and fro at speeds of up to 130mph. Around the world a handful of people may be watching a little more closely … The staff at Dunlop Slazenger … official suppliers of balls to the Wimbledon championships. The women from the factory in Barnsley who have made the balls every year since the second world war. Maybe even the workers in Bataan … who will take over the job for the 2003 tournament. For them, the balls represent … the end result of hours of repetitive hard work, the final act in a meticulously choreographed ballet played out across four continents, and, above all, a livelihood …

'There's a ramshackle air to the Bataan Economic Zone, set up nearly three decades ago by the Philippine government in the hope that western companies would flock in … A huge cinema stands empty; factories abandoned as companies join the search for even cheaper labour in China. Outside the Dunlop Slazenger plant a line of people wait … hoping to bag one of the jobs transferred here from Barnsley.

'Inside the noisy factory are huge machines: high on a platform. [Every few minutes, Ricardo Caranto] opens a hatch and throws a chunk of rubber and a consignment of clay filler or chemicals down a chute into a huge mixer below. Throat-catching dust flies up in his face [and] when a manager asks why he isn't wearing his mask, he says that the factory's dust extraction is pretty good.

'Caranto works shifts, six till two, two till 10, then a 12-hour night shift, and he has been here two years … each week he is moved to a different task so that there is variety. He used to work in a Korean shoe factory but the Dunlop Slazenger plant pays better money: 395 pesos a day – about 65p an hour … Like most of the employees here, [Caranto] is not local. He came here five years ago from Central Luzon, the Philippines' main rice-growing region, to find work. He rents a four-roomed house with his wife and two children half a mile from the factory, and pays 850 pesos (£11) a month in rent.

'Down below him, the machine emits huge blankets of Plasticine-like pink rubber, which pop … as they are fed through metal rollers. Along the way they are installing another … shipped here from Barnsley.

'Cherryle Camara works in the forming room where the balls are finished with two interlocking, elongated ovals of cloth. Several times in a minute she slots two strips of glued material into a clamp, puts the ball in, pulls the clamp over the rubber and removes the near-finished product. It is one of the most difficult jobs – if there's a pucker in the cloth, the ball will be rejected.

'Camara has been here a year and a half. She shares a two-roomed house just up the hill with her parents and her sister, they pay £20 a month. Camara earns about £5 a day' (Abrams, 2002, pp. 2-7).

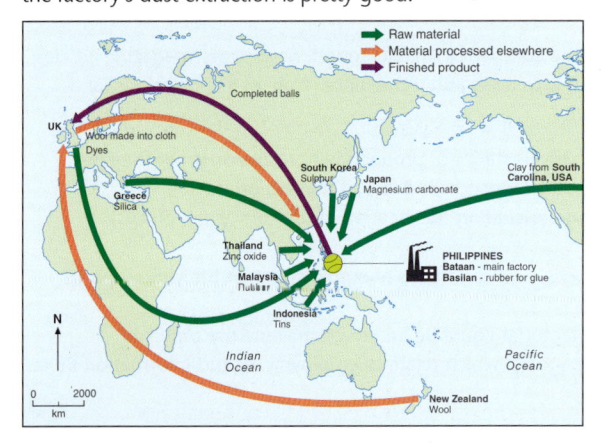

Figure 22: The geography of the manufacture of a Wimbledon tennis ball.

Figure 23: A tennis ball factory worker in Basilan, Philippines. Photo: Sean Smith/Guardian.

CHAPTER 4: GLOBALISATION, STATE AND NATION

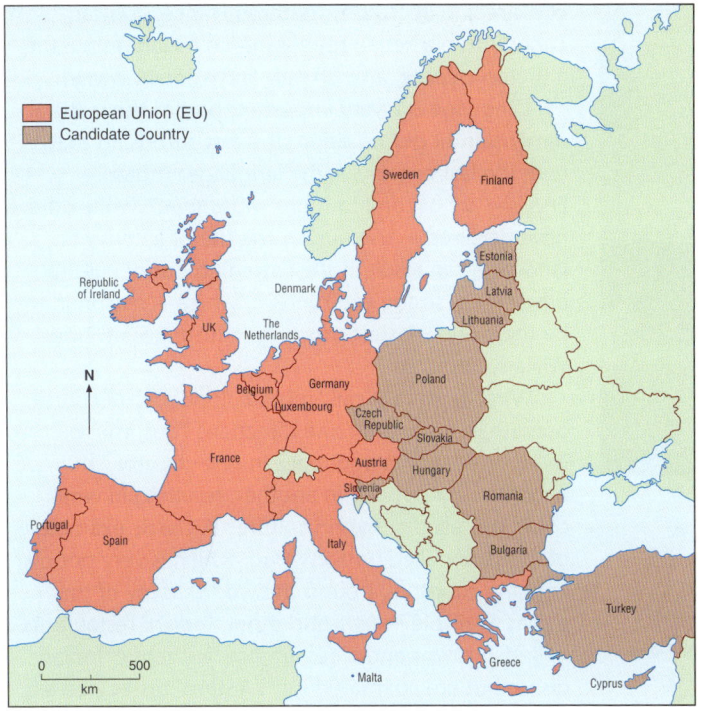

Figure 24: The European Union and candidate countries, 2003.

is seen to be increasingly difficult to cross, particularly for those lacking skills and so, by implication, those from LEDCs. (See pages 9-12).

As well as trading blocs, there are also international institutions which exert considerable influence over the shape of global trade and the economic policies of individual countries. The most prominent of these are the International Monetary Fund and the World Trade Organisation.

Political dimensions
There are those who argue that the ending of the cold war has been a victory for social democracy, and that the ideological divisions of the past no longer exist, or at least have been diminished in their importance. This argument has been popularised through Francis Fukuyama's *The End of History* (1992). The assumption is that with the collapse of communism and the rise of capitalism and liberal democracy, the world will become more politically homogenous. Others, such as Samuel Huntingdon (1996) have, however, suggested that we are moving into an era of a clash of civilisations. Huntingdon sees this in terms of a clash between a Western Christian tradition and an Eastern Islamic one. The events of 11 September 2001 and other similar destructive acts would suggest we are indeed witnessing a military conflict between East and West, with Islamic fundamentalism perceived by many as the successor to communism. However, this is a very simplistic reading of a complex issue. It assumes that the so-called 'Muslim world' is a definable entity in which most people hold one set of views, and it ignores the existence of various kinds of fundamentalism in the West.

At another level, the increasing evidence of political co-operation (e.g. in the EU and between UN member states) is taken as symptomatic of the decline of the importance of single-state action. For example, the struggle against drug trafficking and the so-called 'war against terrorism' can be seen as forms of supra-state policing.

Linked to this, the evolution of systems aimed at harmonising international law, the creation of various international legal protocols, the International Court of Justice in The Hague and a series of related developments, are evidence of the creation of sets of supra-national/state instruments which operate across territorial boundaries and which internationalise politics. A more critical reading of these developments might suggest that some powers, most notably the United States, use these systems selectively – i.e. as and when it suits them (Chomsky, 2000).

There are alternative political visions to those which advocate the continuance of a world state system. For example, the vision of a world free of formalised territorial divisions, where borders do not exist, or a world where the nature and functions of states are built around mutual co-operation rather than competition and conflict. There are those who believe that we should strive for a world in which place of birth and national affiliation are irrelevant, the main goal being one common humanity. Then there are the various movements associated with specific issues such as the environment, or the 'politics of identity' (e.g. gays, lesbians). Such movements operate at all levels from local to international and can be very effective in changing opinions and policy at local, state and even wider scales.

CHAPTER 4: GLOBALISATION, STATE AND NATION

Figure 25: Global tastes in St Tropez, southern France. Photo: David Storey.

Cultural dimensions

There is plenty of evidence around us of a growing 'global culture', and it could be argued that this has supplanted local or national cultures, at least in some aspects. Tastes in music, clothing and food are often cited as examples of this cultural homogenisation (Activity Box 17 and Figure 25). The ubiquitous presence of CNN, MTV and other global media both reflects and sustains this process. This idea of cultural flattening out has been called 'McDonald-isation' or 'Coca-Cola-isation' or just simply 'Westernisation' by its detractors who are critical of the all-pervasive influence of Western capitalism. Figure 25 provides one example of how homogenisation works – along a short stretch of a street in St Tropez, southern France, you can choose from a kebab restaurant, an Italian restaurant, a Breton *crêperie*, an Irish pub and a *sandwicherie*.

Language is a very important part of the cultural dimension of globalisation. It is not surprising that, over the last century, some languages (notably English) are becoming more widespread, while others have declined in use. Just as the use of a common language can result in the flattening out of cultural differences, so too can tourism, which is reckoned to be the world's fastest growing industry (Shaw and Williams, 2002).

What we might call 'cultural flows' do, of course, operate in more than one direction. Thus, music, food, or art which originates in one country can become very popular in another, resulting in what could be called a 'hybrid culture'. A classic example of this is so-called 'Indian' food, which is very popular in many countries around the world. Many of the dishes available in *balti* and other 'Indian' restaurants in British towns and cities are not 'authentic' Indian dishes but are adapted for 'Western' tastes (see e.g. Jackson, 2002). Bhangra music is another example of a cultural form which reflects a fusion of different styles based initially on Punjabi folk music but incorporating many other influences.

Environmental dimensions

Since the 1960s the environment has become the focus of international interest and concern. This is because of growing evidence that the environment is being harmed as a result of human activity, in

---

Activity Box 17: Global cultural products

First, in groups, discuss how the term 'global product' might be defined (think of examples of 'cultural products', e.g. buildings, fashion, art, food, drink).

- In what ways is 'global' different from 'international'?
- What are the main reasons why the term 'global' only came into use in the late twentieth century?
- Think about how some activities, e.g. tourism, can have the effect of 'flattening out' cultural differences. Can you think of any others?

Second, at home take a random selection of about twelve music CDs or records. For each one, try to identify the following:

- Where the performers/composers are from.
- The 'nationality' or place of origin of the style of music. (If this is not possible, try to guess.)
- What other products that you and your family own can be said to be the result of a mixture of styles? Make a list of them.

Compare your results with other members of your group or class.

Information Box 10: Earth Summits

The first 'Earth Summit' was held in Rio de Janeiro in 1992. One of its objectives was to stabilise 'greenhouse gas' concentrations in order to reduce the risk of 'global warming' of climate. A further meeting led to the Kyoto Protocol, which was signed by 39 industrialised countries in 1997. Among other things this introduced an agreement to reduce emissions of six named greenhouse gases by an average of 5.2% from their 1990 levels. The target for this reduction was 2008-12. There has since been active resistance to its implementation in some countries, most notably the United States. The same resistance was evident at the 2002 Earth Summit, held in Johannesburg (see Activity Box 19 for more websites).

One of the main outcomes of the Rio meeting and subsequent Summits was to raise awareness of the need for sustainable development. Progress has been painfully slow and while many countries have adopted the language of sustainability, few have taken major steps to put policy into practice (Huckle and Martin, 2001).

particular increased industrialisation and an increase in vehicle numbers, burning of fossil fuels, dumping of toxic materials in the oceans and on land, and so on. It is now clear that the activities of one country can affect many others (e.g. pollution of the oceans), and that the problem has to be dealt with on a world-wide scale, through international co-operation.

Environmental movements and 'green' politics have helped to raise awareness of the scale of the problem, and since the early 1990s attempts have been made to produce a trans-national blueprint for environmental protection (Information Box 10 and Activity Box 18). Many in the environmental movement believe that the Earth does not 'belong' to anyone, and should be held in trust for future generations.

Summary

As we have seen, there is a growing trend towards multi-state groupings to tackle various issues, and increasing co-operation between states in terms of trade, the environment, and so on. However, while state boundaries may in some ways seem to be 'dissolving', or at least becoming more permeable, they continue to divide the world. As we saw, while MNCs are examples of the 'globalised' economy, they prosper because they are able to take advantage of the different conditions that exist in states around the world, e.g. differences in local wage rates, environmental legislation, tax incentives.

There is an increasing number of non-governmental organisations (NGOs) whose concerns are with issues which affect people or conditions all around the world, regardless of which state they belong to (e.g. health, poverty, gender equality, the environment). Such organisations, and other trans-state movements, help to unite people in common causes. In some ways, then, the idea of 'global citizenship' does exist. However, the forces of globalisation are unlikely to result in the removal of boundaries between states, or obliterate all differences between nations. In other words, we will continue to have multiple identities. Which identity takes priority at any one time will depend on the specific context in which the individual finds him/herself.

Activity Box 18: Sustainable futures

Consult the websites listed below and other sources of reference material to find out more about the Earth Summits.

- The environmental movement often uses the slogan 'Think global, act local'. Explain what you think this means, and give examples of local action that you believe could help to protect the environment on a global scale.
- Why do you think some countries have 'opted out' of agreements to safeguard the environment?
- As a class, or in a group, hold a debate on the subject of 'Economic growth or environmental protection'.

Websites containing information on the Johannesburg 2002 World Summit on Sustainable Development include:

- http://www.johannesburgsummit.org
- http://www.un.org/esa/earthsummit
- http://www.earthsummit2002.org
- http://www.archive.greenpeace.org/search

CHAPTER 5

GLOBAL WORLD, GLOBAL CITIZENSHIP?

A global world?

As we have seen in earlier chapters, the concept of 'globalisation' is not easy to define. Listed below are four aspects of globalisation that we need to consider:

1. Globalisation as a process is not new, it has been happening for centuries.
2. Globalisation does not affect or involve all places, it is spatially uneven.
3. Globalisation does not mean that states will cease to exist or to provide particular functions, or that nations will disappear.
4. There is resistance and opposition to the forces of globalisation.

To take the first point, we know that the process of capital movement and the exchange of particular economic, social and cultural ideas/structures on a global scale has been happening for centuries. For example, from the sixteenth century onwards European powers colonised other parts of the world, resulting in enormous changes in sizeable parts of Africa, Asia and North and South America. The state system (as we currently know it) was, in effect, a European export, as were European languages, some of which came to replace indigenous ones. For example, Spanish is now widely spoken throughout central and South America, while French and English are used in many countries in Africa.

We only need to step inside a museum to see evidence of the fact that ideas, styles, beliefs and modes of behaviour have been interchanged on a worldwide scale for thousands of years. Cultures have always been dynamic and have borrowed heavily from each other. While it may be true that globalisation is happening at a much faster rate than it once did, the phenomenon itself is not new.

The term 'globalisation' itself suggests that it is a process that affects all parts of the world. However, it is a spatially uneven process. Consider, for example, modern forms of communication: there are parts of the world that do not have access to the technologies that many people in MEDCs take for granted. In LEDCs the quality of life for many people is extremely poor – with little or no access to what people in MEDCs think of as 'essential' services, e.g. adequate health care, education, clean water and power (Figure 26). Nevertheless, some of these people may be employees of a multi-national corporation (see Chapters 3 and 4, and Information Box 11, below). (Go to Activity Boxes 19 and 20.)

As well as demonstrating the spatial impacts and unevenness of globalisation, what the above extracts show us is that large corporations benefit from the unevenness of conditions (e.g. labour laws and pollution controls) in different states around the world. They can move operations around the world, shutting down in some locations and opening up in other ones, to take advantage of different regulatory environments.

Figure 26: In Hue, Vietnam, someone has made their home beneath a bridge. Photo: David Storey.

GLOBAL WORLD, GLOBAL CITIZENSHIP?

Information Box 11: How the other half lives

In her book *No Logo*, Naomi Klein describes the working conditions of factory workers in an EPZ (see page 35) in the Philippines. The extract below illustrates how working for a living can itself lead to an early death:

'Zone wages are so low that workers spend most of their pay on shared dorm rooms and transportation; the rest goes on noodles and fried rice from vendors lined up outside the gate. Zone workers cannot dream of affording the consumer goods they produce. These low wages are partly a result of the fierce competition for factories coming from other developing countries ... labor [sic] rights are under such severe assault inside the zones that there is little chance of workers earning enough to adequately feed themselves, let alone stimulate the local economy ... the government views working conditions in the export factories as a matter of foreign trade policy, not a labor-rights issue. And since the government attracted the foreign investors with promises of a cheap and docile workforce, it intends to deliver. For this reason, labor department officials turn a blind eye to violations in the zone or even facilitate them' (Klein, 2000, pp. 210-11).

Klein goes on to tell the story of one worker in this EPZ:

'Carmelita Alonzo ... was a seamstress at the V.T. Fashions factory, stitching clothes for the Gap and Liz Claiborne, among many other labels. [Her] death occurred following a long stretch of overnight shifts during a particularly heavy peak season. "There were a lot of products for ship-out and no one was allowed to go home", recalls Josie [a friend]. "In February [there were] overnights almost every night for one week." Not only had Alonzo been working those shifts, but she had a two-hour commute to get back to her family. Suffering from pneumonia – a common illness in factories that are suffocatingly hot during the day but fill with condensation at night – she asked her manager for time off to recover. She was denied. Alonzo was eventually admitted to hospital, where she died on 8 March 1997 – International Women's Day' (Klein, 2000, p. 216).

Geographies of resistance

There are many arguments against globalisation and these come from different sources and points of view. For example, there are those whose aim is to preserve what they regard as 'uncontaminated' and 'distinct' cultural values which they see as under threat from globalising forces. Such people may blame immigrants or 'foreigners' for changes they do not like – as in the case of white youths blaming Pakistanis in Oldham (see page 13). There are also those who believe that globalisation could lead to homogenisation, and hence the destruction of differences that are of benefit or importance to people, such as local languages, customs and cultural practices. All of these elements are, paradoxically, important to tourism, which is in itself a globalising influence.

In recent years, the anti-globalisation movement has grown around the world, leading to demonstrations and protest campaigns involving various loosely organised groups such as 'Reclaim the Streets'. These groups have protested at meetings of the World Trade Organisation and on their own websites (Figure 27). The 'Jubilee 2000' campaign, which called for the cancellation of Third World debt, involved group protests at the G8 Summit held in Birmingham (15-17 May 1998). Those involved in these movements are opposed to the exploitation of the poor for the profit of big business, and they believe in 'fair trade' (see below).

Minority and group rights

Part of what citizenship is about is respecting difference, and according equal rights to all members of society, regardless of the colour of their skin, their gender, religion, sexual orientation and so on. 'Different but equal' is one way of summing this up. It encourages the idea that difference and diversity are enriching – things to be celebrated rather than feared.

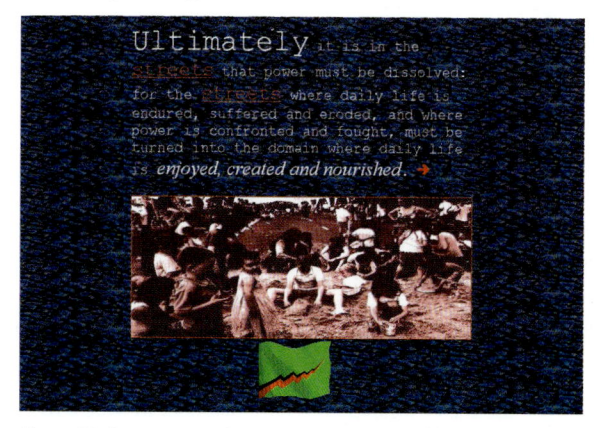

Figure 27: Protests on websites: the Reclaim the Streets statement.

CHAPTER 5: GLOBAL WORLD, GLOBAL CITIZENSHIP?

Figure 28: Gay space in the Castro district of San Francisco. Photo: David Storey.

In any society, we can identify people or groups who face unfair or unequal treatment. In the UK, for instance, these include, but are not restricted to, the following: ethnic minorities, gays and lesbians, the poor, disabled people, the elderly and asylum seekers. Such groups can be excluded from society in various ways: they may be denied access to certain places, facilities, rights, or types of employment that are open to the majority within society; or they may be denied equal respect. The language we use to describe different groups can also be excluding, or discriminatory. For example, in the UK, the term 'Asian' is often used to cover a highly diverse group including people from India, Pakistan, Bangladesh, China, and so on, all of whom have very distinctive cultures and histories. To such people this term is both offensive and inaccurate.

One consequence of unequal treatment or prejudice is that members of 'minority' groups may make their own 'spaces' where they can congregate or live together freely, without fear of disapproval. So, for example, 'gay zones' have developed in many cities, with highly popular and successful bars, clubs and other facilities. In the same way that nations use flags as symbols of their existence, so gay and lesbian communities denote their existence by hanging rainbow flags over the entrances to bars and restaurants (Figure 28). Similarly, 'ethnic' areas have developed in cities, for example, in 'China Town' in New York most of the local population are Chinese, and here shops and restaurants sell and provide Chinese produce.

As suggested earlier, we cannot conceive of citizenship without some consideration of power relationships. People need to be sufficiently empowered to be able to enjoy the rights to which they are nominally entitled. Freedom of movement is of little value if you are too poor to travel. There is also the view that people should be given collective protection from unfair discrimination. Thus, in the UK there is a variety of legislation designed to protect the rights of specified groups. For example, legislation for equal pay for women, employment provisions outlawing various forms of discrimination, anti-racist laws, regulations to allow wheelchair access to buildings.

Activity Box 19: Social spaces

In small groups, think about the social geography of the urban area in which you live or which is nearest to you.

- Can you identify any distinct 'communities' (i.e. where people tend to be similar in terms of, for example, class, ethnicity, age)?
- What evidence did you use to identify these communities?
- Map the locations of particular services related to these communities. Do any patterns emerge? For example, is there any evidence of 'clustering'? If so, try to explain why these clusters occur.

Make use of local newspapers and telephone directories in your research. Look at local community and cultural websites (see, e.g. Chile Sports, Culture and Development Association, Irish Club, Birmingham). Discuss with your teacher the possibility of visiting one of the local community groups or inviting a representative to come and talk about the group's origins, its activities and any links it has with people or similar groups beyond the local area. (Remember the needs of individuals and community groups and you will need to prepare for any such visits with sensitivity.)

Use the information you have gathered to write a report on the presence of a particular community within one part of the urban area. Wherever possible, include geographical, historical and cultural information.

Activity Box 20: Direct consumer action and resistance

Some people argue that we should boycott (i.e. not buy) particular products in order to bring about political change in targeted countries, or to try and change the working practices of a specific multi-national corporation.

- In your opinion, is boycotting products likely to have the desired effect? Give reasons for your answer.
- Would you consider boycotting products that are produced by companies who exploit workers (see e.g. Information Box 11)?
- What alternative strategies might be used?

You could create a 'spoof' poster advertising a particular product that consumers may wish to boycott. The Adbusters website offers an eight-point plan for creating one. It also includes examples of 'spoof' advertisements to inspire you.

Activity Box 21: Investigating fairer trade

Visit the Fairtrade Foundation, Traidcraft and other similar websites (see page 48), and/or research other background information on these organisations.

- Do you think the way they operate is a good idea? Give reasons for your response.
- In the light of what you have discovered, would you consider starting to buy or buying more Fairtrade or Traidcraft products in future?
- Think of ways in which you might try to persuade your friends and family of the benefits of buying fairly-traded products. This could take the form of a poster or leaflet.

Distant others

So far, we have mostly discussed citizenship in the context of individual states/nations, and in terms of our responsibilities towards members of our own society. But what about our responsibilities towards 'distant others', that is people outside and beyond our own patch? Is there such as thing as 'global citizenship'?

For those of us who live in the richer countries of the world, one way in which we can 'think global, act local' is by giving money or goods to world charities and aid agencies who then distribute it to those in need (e.g. to people suffering from famine or the after-effects of natural disasters). However, the showing of 'charitable concern' can sometimes be both patronising and demeaning, however well-intentioned the givers might be. For example, when Christian missionaries went to various LEDCs to convert so-called 'heathens' to Christianity, they were suggesting the superiority of their own traditions and beliefs over those of others.

Another way in which we can 'think global, act local' is through our actions as consumers. As Case Study 3 (page 36) and Information Box 11 (page 41) illustrate, the components of products sold by companies such as Nike and Reebok are quite often manufactured in sweatshop conditions in Thailand, Vietnam or the Philippines. As consumers, we have the power to protest against this type of blatant exploitation – we can stop buying the goods. We can also join campaigns or demonstrate against this type of activity. While governments, such as that in the UK, may complain that anti-globalisation protesters behave in inappropriate and inexcusable ways, we might ask the same question of these large and powerful corporations. Is their behaviour acceptable? (See Activity Box 19.)

Another form of consumer resistance is to purchase only those goods that have been produced under 'fair' employment conditions. Organisations, such as the Fairtrade Foundation and Traidcraft, market such goods (see Activity Box 21). These organisations work to ensure that people in LEDCs obtain fair prices for their products and get a fair share of the benefits of trade. For example, Fairtrade runs a scheme in Dominica, whereby once a fortnight bananas are collected from smallholdings (such as that shown in Figure 29) and shipped to the UK.

Citizenship and the environment

As we saw in Chapter 3, limiting or preventing damage to the natural environment is currently a matter of great concern internationally. It is clear that the consequences of development and

CHAPTER 5: GLOBAL WORLD, GLOBAL CITIZENSHIP?

Figure 29: These fair trade bananas, being boxed at a smallholding in Castle Bruce, Dominica, are destined for the shelves of British supermarkets. Photo: Philip Wolmuth/Panos Pictures.

industrialisation are global in scale. Pollution knows no boundaries, and solutions to pollution problems, whether of water, air or land, require international action and co-operation between governments. What is also needed is a more global mindset on the part of ordinary people; a recognition of the fact that international borders are irrelevant when it comes to global environmental protection.

The concern of most environmental organisations is to protect the environment, and prevent the destruction of natural habitats – both at a local and global level. Figure 30 shows Greenpeace volunteers painting 'toxic crime' on the chimney of a waste incinerator in Sheffield. (According to Greenpeace, the incinerator is the worst in the country for breaking pollution laws – discharging a cocktail of chemicals over the city.

In the last three years it has done so 178 times.) (Go to Activity Box 22.)

The idea of sustainable development is now very popular and widely accepted. This means development which takes account of future needs and is not just based on short-term interests – in other words, saving the planet for the future. Thus, it involves careful use of natural resources as well as economic programmes, which enable development to take place but without damaging environmental consequences.

Citizenship and society

Citizenship as viewed in the context of our own society is generally understood to be about participation and democracy. It is also about citizens behaving in ways that are considered to be 'appropriate' to the welfare and progress of society. According to the current UK government's definition, it is also about being actively involved in the democratic process (Information Box 12). However, not all citizens are active participants in the processes that shape their lives and society, and neither do all citizens have an equal opportunity to participate (as we saw earlier in this chapter). Also, it could be argued that citizenship here is too narrowly defined, and that it should be more to do with human rights than to do with enforcing societal norms.

The UK government clearly wants to encourage us to be actively involved in the political process by using formal channels of expression. However, as the writer George Monbiot (2000) has suggested, active citizenship may also involve a questioning of

Activity Box 22: Environmental pressure groups

Choose one of the well known environmental campaign groups, such as Greenpeace or Friends of the Earth, and find out about it. You should be able to obtain information through websites or direct contact with the organisation.

- What are the aims of the organisation? Does it focus on one environmental issue or on many?
- What strategies does the organisation use to get its message across (e.g. lobbying, protesting, direct action)?
- How can you tell whether the organisation has been successful in protecting the environment?

Would you join a protest such as 'Reclaim the Streets' (see Figure 27) or the 'Jubilee 2000' movement? How would you explain your decision and actions to your friends and family?

GLOBAL WORLD, GLOBAL CITIZENSHIP?

Information Box 12: 'Official' citizenship

The following is an example of the 'official' version of citizenship from the UK government 'Citizenspace' website:

'Get more involved in the democratic process. You can take part in government consultations and discuss views with other users. You can find your elected representatives and get information on elections, or find out how to vote and how to make complaints about public services. Contribute to government policy-making through official consultations, and discuss your views with other users. Explore Citizenspace to see what else you can do.'

Activity Box 23: Definitions of 'good citizenship'?

Using information from the 'Citizenspace' website (see Information Box 12) discuss the following issues as a whole class:

- What do you think the UK government might mean by 'good citizenship'?
- Would you define 'good citizenship' differently? If so, describe how and why.
- Does this concept of 'good citizenship' bear any resemblance to those on other websites?
- Does it cover other issues, such as human rights?

government and the state rather than meekly submitting to actions done in our name or using only 'official' channels to make our views heard (go to Activity Box 23).

Global citizenship

As discussed earlier, citizenship can be seen in terms of our relationships at global as well as local levels – we are all global citizens and therefore have both rights and obligations at a global level. The aid agency Oxfam (see 'Useful websites', page 48) refers to the global citizen as someone who:

- is aware of the wider world;
- respects and values diversity;
- has an understanding of how the world works economically, politically, socially, culturally, technologically and environmentally;
- is outraged by social injustice;
- participates in and contributes to the community at all levels from the local to the global;
- is willing to act to make the world a more equitable and sustainable place; and
- takes responsibility for his or her own actions.

Figure 30: Eleven Greenpeace volunteers occupied the chimney top of the Bernard Road Incinerator in Sheffield for two days in 2003 to prevent the incinerator being used. Photo: ©Greenpeace/Sims.

CHAPTER 5: GLOBAL WORLD, GLOBAL CITIZENSHIP?

Activity Box 24: Becoming a global citizen

Consider the two lists above and all that you have learnt from this book.

- Do you mostly agree or disagree with Oxfam's choice of criteria for what defines a 'global citizen'?
- Would you add or take away any of the criteria, and if so why?
- For each item in the Oxfam list, give two examples of the kind of action that you could or already do take to qualify as a 'global citizen'.
- Consider the points that describe 'an effective citizen'. Do any of the characteristics match any of the UK government's statement on being a good citizen? If so, how? If not, in what ways do they differ?
- Rank the eight characteristics of an effective citizen in order of importance to you as an individual.
- Explain why you believe that some are more important than others and provide examples to illustrate your argument.

When consulted about what defines an 'effective citizen', experts from nine nations reached a consensus on the eight characteristics listed below. An effective citizen is someone who:

1. looks at problems in a global context;
2. works co-operatively and responsibly;
3. accepts cultural differences;
4. thinks in a critical and systematic way;
5. solves conflicts non-violently;
6. changes lifestyles to protect the environment;
7. defends human rights;
8. participates in politics.

(Cogan and Derricott, cited in Hicks, 2001, p. 57)
(Go to Activity Box 24.)

Geography matters

While we have understandable attachments to our own places and countries, in the twenty-first century it appears appropriate to develop a citizenship, which combines a knowledge of the local with an empathy for the world beyond our own immediate horizons. Only by learning more about others' lives, politics and environments can we develop such an empathy, and on that basis work towards an understanding, peaceful, tolerant and just future as global citizens.

REFERENCES AND FURTHER READING

References

Abrams, F. (2002) 'New balls, please', *Guardian G2*, 24 June, pp. 2-7.

Aitchison, J. and Carter, H. (1994) *A Geography of the Welsh Language*, 1961-1991. Cardiff: University of Wales Press.

Amnesty International (1997) 'How is a refugee is made?' (http://www.refuge.amnesty.org/htm/how2.htm), accessed 10 February 2003.

Anderson, B. (1991) *Imagined Communities. Reflections on the origin and spread of nationalism* (revised edition). London: Verso.

August, O. (2000) *Along the Wall and Watchtowers. A journey down Germany's divide*. London: Flamingo.

Bell, D. and Valentine, G. (eds) (1995) *Mapping Desire. Geographies of sexualities*. London: Routledge.

Billig, M. (1995) *Banal Nationalism*. London: Sage.

BNP (2003) 'Immigration: Time to say no!' (http://www.bnp.org.uk/policies.html#immigration), accessed 10 February.

Chatwin, B. (1998) *On the Black Hill*. London: Vintage Classics.

Chomsky, N. (2000) *Rogue States. The rule of force in world affairs*. London: Pluto Press.

Collins, R. (1990) *The Basques* (second edition). Oxford: Blackwell.

Dicken, P. (1998) *Global Shift. Transforming the world economy* (third edition). London: Paul Chapman.

Earle, R. and Davies, D. (1998) *One Love. The Reggae Boyz: An incredible soccer journey*. London: André Deutsch.

Faulks, K. (2000) *Citizenship*. London: Routledge.

Fonseca, I. (1996) *Bury Me Standing*. London: Vintage.

Fukuyama, F. (1992) *The End of History and The Last Man*. New York: Free Press.

Foucault, M. (1980) *Power/Knowledge. Selected interviews and other writings* (edited by Gordon, C.). Brighton: Harvester Press.

Gilroy, P. (1987) *'There Ain't no Black in the Union Jack': The cultural politics of race and nation*. London: Hutchinson.

Gramsci, A. (1971) *Selections from the Prison Notebooks*. New York: International Publishers.

Harvey, D. (1989) *The Condition of Postmodernity*. Oxford: Blackwell.

Haywood, P. (ed) (2002) *Poems for Refugees*. London: Vintage.

Heffernan, M. (1998) *The Meaning of Europe: Geography and geopolitics*. London: Arnold.

Held, D., McGrew, A., Goldblatt, D and Perraton, J. (1999) *Global Transformations. Politics, economics and culture*. Cambridge: Polity Press.

Hicks, D. (2001) 'Envisioning a better world', *Teaching Geography*, 26, 2, pp. 57-60.

Holmes, M. and Storey, D. (forthcoming) 'Who are the boys in green? Irish identity and soccer in the Republic of Ireland' in Porter, D. and Smith, A. (eds) *Sport and National Identity in the Post-war World*. London: Routledge.

Huckle, J. and Martin, A. (2001) *Environments in a Changing World*. Harlow: Prentice Hall.

Huntingdon, S.P. (1996) *The Clash of Civilizations and the Remaking of World Order*. New York: Simon and Schuster.

Jackson, P. (1989) *Maps of Meaning. An introduction to cultural geography*. London: Routledge.

Jackson, P. (2002) 'Geographies of difference and diversity', *Geography*, 87, 4, pp. 316-23.

Kinsman, P. (1995) 'Landscape, race and national identity: the photography of Ingrid Pollard', *Area*, 27, 4, pp. 300-10.

Kitchin, R. (2000) *Changing Geography: Disability, space and society*. Sheffield: Geographical Association.

Klein, N. (2000) *No Logo*. London. Flamingo.

Kurlansky, M. (2001) *The Basque History of the World*. London: Penguin.

McDowell, L. and Sharp, J. (eds) (1997) *Space, Gender, Knowledge. Feminist readings*. London: Arnold

Marshall, T.H. (1950) *Citizenship and Social Class*. Cambridge: Cambridge University Press.

Massey, D. (1994) *Space, Place and Gender*. Cambridge: Polity Press.

Miller, D. (1997) *On Nationality*. Oxford: Clarendon Press.

Monbiot, G. (2000) *Captive State. The corporate takeover of Britain*. London: Macmillan.

Monger, C. (1995) *The Englishman Who Went Up a Hill but Came Down a Mountain*. London: Corgi.

Morton, H.V. (2000) *In Search of England*. London: Methuen.

Motion, A. (ed) (2001) *Here to Eternity: An anthology of poetry*. London: Faber and Faber.

Shaw, G. and Williams, A.M. (2002) *Critical Issues in Tourism. A geographical perspective* (second edition). Oxford: Blackwell.

Sibley, D. (1995) *Geographies of Exclusion. Society and difference in the West*. London: Routledge.

Smith, A.D. (1991) *National Identity.* London: Penguin, London.

Smith, D.M. (ed.) (1992) *The Apartheid City and Beyond. Urbanization and social change in South Africa.* London: Routledge.

Storey, D. (2002) 'Territory and national identity: examples from the former Yugoslavia', *Geography*, 87, 2, pp. 108-15.

Taylor, P.J. and Flint, C. (2000) *Political Geography. World-economy, nation-state and locality* (fourth edition). Harlow: Prentice Hall.

Unwin, T. (1999) 'Place, territory and national identity in Estonia' in Herb, G.H. and Kaplan, D.H. (eds) *Nested Identities. Nationalism, territory and state.* Lanham: Rowman and Littlefield, pp. 151-73.

Yarwood, R. (2002) *Changing Geography: Countryside conflicts.* Sheffield: Geographical Association.

Further reading

Cloke, P., Crang, P. and Goodwin, M. (eds) (1999) *Introducing Human Geographies.* London: Arnold.

Cogan, J. and Derricott, R. (1998) *Citizenship for the 21st century.* London: Kogan Page.

Holloway, L. and Hubbard, P. (2001) *People and Place. The extraordinary geographies of everyday life.* Harlow: Prentice Hall.

Miliband, R. (1969) *The State in Capitalist Society.* London: Quartet.

Smith, D. (1999) *The State of the World Atlas* (sixth edition). London: Penguin.

Storey, D. (2001) *Territory. The claiming of space.* Harlow: Prentice Hall.

Sutcliffe, B. (2001) *100 Ways of Seeing an Unequal World.* London: Zed Books.

Valentine, G. (2001) *Social Geography. Space and Society.* Harlow: Prentice Hall.

Useful websites

- Adbusters – http://www.adbusters.org (anti-consumerist site, famous for 'spoof' advertisements)
- Amnesty International – http://www.amnesty.org/
- British National Party – http://www.bnp.org.uk
- Chile Sports, Culture and Development Association, Sheffield – http://www.chilescda.org.uk/
- Council for the Protection of Rural England – http://www.cpre.org.uk/main.htm
- Countryside Alliance – http://www.countryside-alliance.org/
- Czech Centre – http://www.czechcenter.com/ROMA.htm
- Earth Summit 2002 – http:www.earthsummit2002.org
- Fair Trade – http://fairtraderesource.org/
- Greenpeace – http://www.archive.greenpeace.org/search
- Helsinki Citizens Assembly – http://www.czechia.com/hcaroma/default.htm
- Home Office (UK government) – http://www.homeoffice.gov.uk/rds/statsprog1.html
- Human Rights Watch – http://hrw.org/reports/world/czech-pubs.php
- Immigration and Nationality Directorate (UK) – www.ind.homeoffice.gov.uk/
- Immigration and Naturalisation Service (US) – http://www.immigration.gov/graphics/services/natz/general.htm
- Infoplease – http://www.infoplease.com/ipa/A0107447.html
- Ingrid Pollard's photographs and text – http://www.autograph-abp.co.uk/gallery/pol.html
- Irish Club, Birmingham – http://www.icirishclub.co.uk/
- Johannesburg Earth Summit – http://www.Johannesburgsummit.org
- No Logo – http://www.nologo.org
- Oxfam – http://www.oxfam.org/
- Phatfotos – http://www.phatfotos.com/carnival
- Reclaim the Streets – http://www.reclaimthestreets.net/ and http://rts.gn.apc.org/prop01.htm
- Refugee Council – http://www.refugeecouncil.org.uk/
- Roma in the Czech Republic – http://www.romove.cz/romove/situation.html
- Traidcraft – http://www.traidcraft.org/
- Treasures of the World – http://www.pbs.org/treasuresoftheworld/a_nav/guernica_nav/main_guerfrm.html
- Citizenspace – http://www.gov.uk/online/citizenspace/default.asp
- United Nations – http://www.un.org/ and Declaration http://www.un.org/Overview/rights.html
- United Nations Earth Summit – http://un.org/esa/earthsummit
- University College Worcester – http://www.worc.ac.uk